Dawn Cole

STRATEGIC COST REDUCTION

Leading Your Hospital to Success

American College of Healthcare Executives Management Series Editorial Board

Alan M. Zuckerman, FACHE
Health Strategies & Solutions, Inc., Philadelphia, PA

Lynn I. Boggs, FACHE
Presbyterian Health/Novant, Charlotte, NC

Ralph Charlip, FACHE
VA Health Administration Center, Denver, CO

Terence T. Cunningham, III, FACHE
Ben Taub General Hospital, Houston, TX

Dennis E Franks, FACHE
Neosho Memorial Regional Medical Center, Chanute, KS

Nick Macchione, CHE
San Diego Health & Human Services Agency, San Diego, CA

Kimberly K. Sharp
HCA Nashville, TN

John M Snyder, FACHE
Carle Foundation Hospital, Urbana, IL

CDR Mark J Stevenson, FACHE
TRICARE Regional Office–West Aurora, CO

Mikki K Stier, FACHE
Broadlawns Medical Center, Des Moines, IA

Emily L. Young, CHE
Central Mississippi Medical Center, Jackson, MS

Fredrick C. Young, Jr., FACHE
Methodist Health System Foundation, Slidell, LA

STRATEGIC COST REDUCTION

Leading Your Hospital to Success

Michael E. Rindler

ACHE Management Series
Health Administration Press

Your board, staff, or clients may also benefit from this book's insight. For more information on quantity discounts, contact the Health Administration Press Marketing Manager at (312) 424-9470.

This publication is intended to provide accurate and authoritative information in regard to the subject matter covered. It is sold, or otherwise provided, with the understanding that the publisher is not engaged in rendering professional services. If professional advice or other expert assistance is required, the services of a competent professional should be sought.

The statements and opinions contained in this book are strictly those of the author and do not represent the official positions of the American College of Healthcare Executives or of the Foundation of the American College of Healthcare Executives.

Copyright © 2007 by the Foundation of the American College of Healthcare Executives. Printed in the United States of America. All rights reserved. This book or parts thereof may not be reproduced in any form without written permission of the publisher.

11 10 09 08 07   5 4 3 2 1

Library of Congress Cataloging-in-Publication Data

Rindler, Michael E.
 Strategic cost reduction: leading your hospital to success / Michael E. Rindler.
    p. cm.
 Includes bibliographical references.
 ISBN-13: 978-1-56793-267-6 (alk. paper)
 ISBN-10: 1-56793-267-3 (alk. paper)
 1. Hospitals—Finance. 2. Hospitals—Buisness management. I. Title.
 RA971.3.R565 2007
 530.4'15—dc22
                                                            2006051011

The paper used in this publication meets the minimum requirements of American National Standard for Information Sciences—Permanence of Paper for Printed Library Materials, ANSI Z39.48-1984. ∞ ™

Acquisitions editor: Janet Davis; Project manager: Amanda Bove; Layout editor: Chris Underdown; Cover designer: Trisha Lartz

Health Administration Press
A division of the Foundation of the
  American College of Healthcare Executives
1 North Franklin Street, Suite 1700
Chicago, IL 60606-3424
(312) 424-2800

Contents

Acknowledgments — vii

Foreword — ix

Preface — xiii

PART I: THE STRATEGIC COST REDUCTION IMPERATIVE — 1

Chapter 1	What Is Strategic Cost Reduction, and Why Is It So Important?	3
Chapter 2	What Does It Take to Succeed?	15
Chapter 3	Why Do Some Hospitals Fail?	25
Chapter 4	Extraordinary Success Story at Northeast Georgia Medical Center	33

PART II: BEST OPPORTUNITIES FOR STRATEGIC COST REDUCTIONS — 45

Chapter 5	Management Costs	47
Chapter 6	Labor Costs	55
Chapter 7	Supply Costs	65
Chapter 8	Service Costs	73
Chapter 9	Clinical Utilization Costs	83
Chapter 10	Capital Costs	91

Part III: Practical Approaches to Strategic Cost Reduction — 97

Chapter 11	Setting the Stage for Success	99
Chapter 12	120 Days to Strategic Cost Reduction Success	109
Chapter 13	Internal Approach	127
Chapter 14	Consultant Approach	133
Chapter 15	Pitfalls to Avoid	141

Part IV: Success Through Implementation — 147

Chapter 16	Monitoring Strategic Cost Reductions	149
Chapter 17	Accountability for Strategic Cost Reduction	157
Chapter 18	Building a Winning Strategic Cost Reduction Culture	163

Afterword — 169

Appendix A: Background Information Trend Guide — 171

Appendix B: Sample Calendar for Strategic Cost Reduction — 173

Appendix C: Sample Strategic Cost Reduction Scenario — 179

Appendix D: Sample Town Meeting Questions and Answers — 181

Appendix E: Sample Task Force Guidelines — 185

Appendix F: Sample Task Force Meeting Guidelines — 189

Appendix G: Sample Weekly Task Force Report — 191

Appendix H: Sample Board of Directors Report — 193

Appendix I: Sample Detailed Cost Reduction — 195

About the Author — 197

Acknowledgments

Janet Davis, acquisitions editor at Health Administration Press, deserves full credit for planting the idea that eventually grew into this book. Her enthusiasm and guidance are very much appreciated.

I would like to thank my colleague and friend, Dan Beckham, president, The Beckham Company, for introducing me to wonderful clients like Northeast Georgia Medical Center. The Medical Center undertook a strategic turnaround that is succeeding beyond all expectations. The simultaneous completion of a new strategic plan under Dan Beckham's guidance and a strategic cost reduction with my assistance allowed a powerful leadership team—composed of board, administration, medical staff, and management—to achieve great success in their Gainesville, Georgia, marketplace.

Finally, I would like to acknowledge Leslie Rindler and Megan DeWalt. Leslie prepared countless drafts of the manuscript. Megan provided expert editorial and copy assistance. Together, we are proud of the final result, and we hope *Strategic Cost Reduction* will make a meaningful contribution to individual hospital leadership teams and will have a positive effect on reducing hospital costs nationwide.

Foreword

THE FIRST OBLIGATION of a chief executive officer (CEO) is to ensure the continued vitality of the enterprise so that it can accomplish its mission. For a hospital, health system, or physician organization, that mission is clear-cut: It is to improve health. The path to this mission is not clear-cut. Indeed, it is one of the most complex of all modern management challenges, as preeminent management advisor Peter Drucker and others have observed. Many factors contribute to this complexity, but one of the most important is rapid and continuous advances in the methods used to deliver care. This reality requires healthcare executives to invest in technology and human capabilities at a level that is unprecedented in other industries. The capital for such investment can come from only two places: It can be borrowed, or it can be generated from operations. There are only two levers you can pull when it comes to generating capital from operations: You can increase the volume of profitable services you provide, or you can reduce your costs.

Thinking about cost reduction along a continuum of relative importance is helpful. At one end of the continuum is the kind of daily effort that goes into keeping costs low—for example, buying the less costly catheter when quality is equivalent. At the other end of the continuum is what can best be described as strategic cost reduction. This is the kind of cost reduction an organization undertakes when the stakes are high. The stakes are high when the enterprise's ability to access capital to invest in technology and human capabilities is challenged.

Cost reduction becomes strategic when it has one or more of these key characteristics:

- It cannot be achieved unless it is driven by the CEO.
- It has become an item of discussion on the board agenda.
- It requires engagement of the entire organization.
- It must produce a significant shift in operating margin.
- It is necessary to achieve the organization's strategic plan.

Strategic cost reduction requires an approach that is comprehensive and disciplined. This approach must produce significant results in the short term while setting in place the mechanisms that ensure that the results can be sustained in the long term. More than anything, sustained results require the commitment of individuals at all levels of the organization who understand and support the importance of generating the margins necessary to buy new technology, recruit physicians and staff, build and improve facilities, invest in training, and provide attractive compensation and benefits. In short, it requires employees, physicians, lenders, and board members who are committed to not falling behind in the race to secure an advantageous position in a rapidly transforming environment. The organization's playbook for achieving that advantageous position is its strategic plan. Strategic planning is allocating scarce resources to your best opportunities. Strategic cost reduction makes scarce resources less scarce.

In its most powerful form, strategic cost reduction proceeds within the broader context of a comprehensive strategic planning process that provides the clear rationale for why cost reduction is essential to the organization's future. It is possible and often desirable for the strategic planning process and the strategic cost reduction process to proceed simultaneously. This simultaneous approach has three benefits. First, it clearly connects cost reduction to the organization's vision and driving strategies. Second, it engages the organization in a single push rather than trying to command attention for the strategic plan first and then trying to command its attention again in a second round of engagement focused on cost reduction. Finally, it produces a tangible and significant result for the strategic plan that has meaning organization-wide. Too many strategic plans remain distant and theoretical documents with a decidedly fuzzy connection to the real work of the organization.

An approach that clearly connects cost reduction with the aspirations of the organization, as set forth in its strategic plan, produces surprising levels of commitment. Rather than regard the cost reduction effort as a threatening, top-down fiat that is done to the organization, it comes to be viewed as an investment in securing a prosperous and healthy future for employees, physicians, and the community. As a result, physicians, department directors, and first-line employees become advocates rather than victims. Cost cutting for the sake of cost cutting can become both heartless and headless: Heartless because it can proceed without sufficient concern for its human consequences, and headless because, without forethought, it can end up cutting into those capabilities on which the organization's future is most likely to depend.

Strategic cost reduction is a cornerstone in a strategic turnaround. In addition to cost reduction, a strategic turnaround depends on market-share growth in profitable service lines and development of a vital, balanced, and committed medical staff. Evidence of a meaningful strategic turnaround is demonstrated by significant improvement in three categories of performance:

1. Operating margins
2. Reputation
3. Morale and commitment among employees and physicians

Alan Bleyer, FACHE, president and CEO of Akron General Health System in Akron, Ohio, used a strategic planning process as the catalyst for producing a significant turnaround that stemmed several years of multimillion dollar losses and produced dramatic marketshare growth. Fred Loop, M.D., did the same thing at The Cleveland Clinic. When Jim Gardner took over as CEO at Northeast Georgia Medical Center in Gainesville, Georgia, he wasn't faced with the kind of losses Bleyer and Loop had faced, but he quickly determined that the organization's financial performance was too anemic to fund the investment necessary to keep the organization from falling behind aggressive and expansive competitors in Atlanta. Chapter 4 of this book profiles Gardner's efforts to rapidly move his organization into a dramatically improved position related to margins, reputation, morale, and commitment. It was Gardner who decided, despite some skepticism, that Northeast Georgia could undertake a strategic planning process and a cost reduction effort simultaneously. And he was the one who kept this effort balanced and moving forward.

Michael Rindler worked closely with Gardner and helped him design and apply the approach described in this book. He has developed and applied this approach effectively with numerous hospitals and healthcare systems. Michael understands the economic engine of a healthcare organization, and he knows how to tune it for significant improvements in performance. His approach works.

—J. Daniel Beckham, President
The Beckham Company
Blufton, South Carolina

Preface

EVERY HOSPITAL HAS the ability to reduce its total costs by at least 5 percent annually without harming clinical quality or customer service. How do I dare make this statement? Can it be true? If it were true, what impact could it have on our troubled healthcare system? Strategic cost reduction is the pathway to this extraordinary achievement. Its impact is profound.

In my 30-year career as a hospital chief executive officer (CEO) and healthcare strategic adviser, I have helped hospitals reduce annual costs by about a quarter of a billion dollars. Some of these cost reductions were part of turnarounds I led for highly distressed hospitals that were seeking only to survive. For the past decade, I have advised a number of already successful hospitals and health systems on how to strategically reduce costs to further enhance overall performance. The insights gained from these strategic advisory assignments, which ranged from small rural hospitals with $30 million in annual revenues to urban academic medical centers with $1 billion in annual revenues, form the principal messages of this book. These messages are universally applicable to all hospitals.

Throughout the United States, finance and cost issues are the number-one concern of hospital CEOs. Medicare and Medicaid reimbursement systems are strained to the limit, and no relief is in sight. Years of focusing on revenue maximization, volume growth, and managed-care strategies have not left hospitals noticeably better off. It is my belief that strategically reducing costs should be the next highest-priority leadership challenge for hospitals of all sizes and in all locations.

Use of the phrase "leadership challenge" is not casual. I do not view strategic cost reduction as a financial challenge. Finance officers deserve neither the exclusive credit when hospitals do well financially nor the exclusive blame when hospitals perform poorly. Finance officers review and record performance. They also provide valuable advice and guidance to the operating managers within the hospital. In my judgment, however, finance officers play a supporting role in strategic cost reduction, albeit an important one.

This book is addressed to hospital CEOs, senior management (chief operating officers, vice presidents, and assistant administrators), and middle management (directors, managers, and supervisors). These are the people who produce the hospital's revenue and spend the hospital's resources. This book should also be very useful to boards of directors and physician leaders.

By adopting the essential principles of strategic cost reduction, individual hospitals and our nation's healthcare system can tremendously improve the way resources are spent. What are these principles?

1. Hospital senior management teams must have the fundamental knowledge of where the money goes; my moniker for this is "green dollars." Green dollars are checks written (or electronic transfers made) to pay salaries and fringe benefits and to purchase supplies and services for patient care and the myriad of other activities that support patient care within the hospital.
2. Senior management teams must use the knowledge that middle managers, physicians, and employees have, as they are closest to where green dollars are being spent.
3. Senior management teams must trust what the middle managers, physicians, and employees recommend for reducing costs without compromising quality and service.

4. CEOs and senior management teams must create a corporate culture of competence and set a highly positive example for middle management, physicians, and employees.

These four principles can work flawlessly to strategically reduce hospital costs to the lowest possible levels without compromising high clinical quality and excellent customer service. Strategic cost reduction is not the only approach to reducing hospitals, but it is by far the best approach. Other approaches are short-term, stop-gap measures at best. Strategic cost reduction enables hospitals to achieve long-term, sustainable results.

Although this is a book about strategic cost reduction, its content is mostly about leadership techniques rather than financial techniques. Instead of a detailed prescription for making cost reduction decisions, this book describes a type of leadership that gives great importance to the presence of sound cost reduction ideals. While hospitals differ in size, geographic location, and complexity of services, they all have middle management, physicians, and employees. This book is based on the wisdom of asking these three groups for opinions about strategic cost reductions and trusting that constructive ideas will be offered. If a hospital's CEO and senior management have first established a winning leadership culture, universal success with strategic cost reduction will follow.

This book provides the pathways to success for hospital leadership teams who are committed enough, capable enough, and focused enough to take them. Can it be that simple? Yes. In fact, it is stunningly simple: The essence of strategic cost reduction success is good leadership.

PART I

The Strategic Cost Reduction Imperative

HOSPITAL COSTS ARE too high. Yet reducing hospital costs takes leadership vision, commitment, and discipline. There are no easy strategies and no magic bullets leading to instant success.

Chapters 1 through 4 define the strategic cost reduction imperative. Beginning with my definition in Chapter 1, strategic cost reduction is highlighted as a hospital leadership challenge, not merely a financial challenge. Chapter 2 articulates key factors for achieving success through strategic cost reduction, and Chapter 3 covers common reasons why some hospitals fail. Chapter 4 is a case study of an extraordinary strategic cost reduction success for a Georgia health system, which offers valuable lessons for all hospitals, regardless of size or geographic location.

The strategic cost reduction imperative is underscored by professional challenges hospitals face today and in the foreseeable future. Hospital chief executive officers (CEOs) place financial issues at the very top of their list of concerns. Part I places a proactive leadership solution to financial concerns in the hands of CEOs and their senior management teams. Strategic cost reduction is one solution to this nation's ongoing healthcare financing crisis. Strategic cost reduction is within the means of every hospital where the leadership team has courage, competence, and commitment.

CHAPTER 1

What Is Strategic Cost Reduction, and Why Is It So Important?

ALL HOSPITALS SPEND money, regardless of their size, prestige, or geographic location. Some spend efficiently, and others waste money that could have been put to better use caring for patients. A common theme is found among those who spend most efficiently. Hospitals that take a strategic view toward costs aspire to achieve the lowest costs without compromising excellent clinical quality and patient service. Strategic cost reduction is a way of life for these successful hospitals.

Strategic cost reduction success has its foundation in extraordinary leadership quality. Strategic cost reduction cannot be achieved without the firm and full commitment of a hospital's CEO and senior management team (vice presidents or assistant administrators, and the CEO's direct reports). Although the phrase "strategic cost reduction" may seem to imply an emphasis on accounting and finance, strategic cost reduction success cannot be achieved by focusing on dollars alone. Excellent leadership is what really makes it work.

WHAT IS STRATEGIC COST REDUCTION?

Strategic cost reduction is defined as reducing costs to the lowest possible levels without harming quality clinical care and customer service, while achieving the hospital's strategic goals. Strategic cost reduction requires a hospital to have a strategic plan with long-term strategic goals. Strategic cost reduction is a process of carefully controlling costs to achieve a level of financial performance that enables execution of the hospital's long-term strategic goals.

Strategic cost reduction requires a leadership culture that simultaneously appreciates efficiency and low costs while highlighting excellent quality and customer service. The creation of this special leadership culture is an absolute necessity for strategic cost reduction success. It takes a highly capable leadership team—composed of the CEO and an excellent senior and middle management staff—to attain this level of success. It takes a creative leadership team to consider alternatives in staffing models, supply purchases, and clinical utilization to achieve the same positive patient care outcome. It takes an efficient leadership team to select the best possible resources, not necessarily the most expensive resources. It takes a persistent leadership team to continuously reevaluate financial and patient care results and to use that feedback to achieve cost reduction success.

WHY IS STRATEGIC COST REDUCTION SO IMPORTANT?

Achieving strategic cost reduction success allows a hospital to maximize its value for its customers—that is, its patients. Achieving the lowest possible costs allows a hospital to simultaneously attain the highest possible quality and the best customer service. Achieving strategic cost reduction success takes leadership discipline; there are no easy techniques or shortcuts to success. Accountability for achieving excellent strategic cost reduction results rests with the CEO and the senior management team. It cannot be delegated to others, whether they are hospital consultants or outside consultants.

For hospitals, strategic cost reduction success is important because it means having the highest possible employee productivity.

It means being efficient purchasers of supplies and services. It means reaching optimal levels of clinical utilization—in other words, striking a balance between too little utilization of clinical services, which can lower quality, and too much utilization, which can waste precious healthcare dollars.

GREEN DOLLARS

Strategic cost reduction success requires a complete understanding of the concept of "green dollars." Green dollars are the dollars expended to achieve the hospital's purpose. They include salary and benefit costs as well as the costs to purchase the supplies and services required to provide patient care and support services. To achieve strategic cost reduction success, green-dollar costs must be evaluated and reduced wherever possible. A hospital does not attain strategic cost reduction success by merely attempting to reduce budgets. A hospital that truly understands the concept of green dollars knows it must reduce the actual payroll dollars, supply dollars, and service dollars to achieve real strategic cost savings.

Oftentimes, hospitals fool themselves into thinking they are reducing costs. For example, eliminating an unfilled or budgeted staff position may sound great, but it does not actually reduce green-dollar labor costs. To achieve strategic cost reduction success, the hospital has to stop writing checks for employee salaries, fringe benefits, supplies, and services. A check not written translates into green dollars saved. Many hospitals fail to achieve the management discipline required to eliminate green-dollar expenses. Those hospitals more often than not fool themselves year after year, until the board of directors or lenders lose patience. When that happens, senior management should be replaced by people who better understand the concept of green dollars and are better equipped to achieve strategic cost reduction success.

THE STRATEGIC COST REDUCTION LEADERSHIP IMPERATIVE

Strategic cost reduction leads to improved financial performance for any hospital. Better financial performance ultimately leads to better access to capital, which leads to expansion and improvement in the hospital's services. Optimum financial performance creates a corporate culture in which the best physicians and the best employees wish to work. In these instances, strategically lowering costs can and will lead to the hospital's higher overall performance. This higher performance translates into more satisfied patients and, essentially, to more customers and volume growth for hospital services.

Hospital leaders who understand and successfully practice strategic cost reduction can succeed in almost any environment, regardless of how challenging. For example, there are inner-city hospitals that understand strategic cost reduction and survive against all odds. These hospital leaders do not whine about how terrible the hospital's reimbursement is or how difficult their state's regulatory environment may be. Instead, they focus their leadership talents on reducing costs effectively, wherever possible, and finding ways to thrive under adversity. Learning to cope with lower reimbursement rates, rather than complaining about them, helps maximize the hospital's effectiveness amid the most challenging reimbursement situations.

Against the odds

An urban hospital faced numerous challenges that threatened its survival and created a potential loss of much-needed hospital services for a highly disadvantaged inner-city neighborhood. Years of operating in the red and caring for uninsured patients had depleted the hospital's financial reserves as well as physician and employee morale.

> The hospital's CEO confronted these seemingly insurmountable challenges with a highly positive attitude. He candidly shared the hospital's gloomy financial picture with physicians and employees. He told them bluntly that they would have to make profound operating cost cuts to survive. He challenged physicians and employees to rethink all expenditures with a fresh perspective and a positive vision of the future. He truly inspired the hospital's management, physician, and employee constituencies to create a new way of operating the hospital that would allow it to remain within its limited financial means.
>
> The results of this strategic cost reduction approach were dramatic. Through a combination of labor, fringe benefit, supply, and service cost reductions, green-dollar hospital costs were reduced 13 percent overall. Physicians made dramatic improvements in clinical utilization practices, notably dropping inpatient length of stay by 20 percent. The first full year after implementing these cost reduction initiatives became the hospital's first profitable year in more than a decade. The hospital's senior and middle management, along with physician leaders, proved they could survive against all odds by using strategic cost reduction to revitalize what had been a dying inner-city hospital.

Hospitals situated in the most affluent locations and the most favorable reimbursement climates that do not understand strategic cost reduction will ultimately fail to thrive. Not understanding the concept of strategic cost reduction often results in the waste of precious resources. What is the difference between hospital success and failure? It is leadership. Great hospital leaders will succeed because they can find ways for their organizations to thrive in the most adverse conditions. For these hospital leaders, strategic cost reduction is a leadership imperative.

OTHER COST REDUCTION APPROACHES

Hospitals use many approaches to reduce costs; some are more successful than others. Strategic cost reduction is one approach. What about others?

- *Turnaround cost reduction efforts*, usually guided by outside consulting companies, are used for financially challenged hospitals. These efforts involve employee layoffs, reductions in programs or clinical services, and massive changes in supply and service purchasing practices.
- *Budget cutting* is another cost reduction technique, routinely deployed when the following year's financial projections do not equal desired levels of future performance.
- *Across-the-board cost cutting* is used to reduce costs by a predetermined percentage. This approach to cost cutting is usually used in hospitals where senior management is in trouble and needs to improve financial performance immediately.
- *Benchmarking* is another technique hospitals use to cut labor or other costs to match the performance of similar hospitals.
- *Supply chain management*, which was popularized by hospital consulting companies, is yet another cost reduction technique focused on reducing supply costs.

All of these approaches have merit; however, none of them is as effective as strategic cost reduction. Only strategic cost reduction ensures long-term success. This success requires more commitment, more personal involvement, and more leadership than other, less-effective approaches.

LENDER PERSPECTIVES AND CAPITAL ACCESS

Hospitals have many sources of investment capital. Not-for-profit hospitals have access to nonprofit bond funding if their financial performance meets minimum criteria. Banks and other lenders are sources of capital for both excellently performing and poorly performing organizations. The difference, of course, is accounted for when the banks and lenders charge higher interest rates for poorly performing hospitals. Lenders evaluate the risk profiles of hospitals by evaluating many parameters. Financial performance is one of the most important. Hospitals able to achieve the lowest possible cost structures, regardless of their reimbursement environment, achieve higher operating margins. Higher-cost hospitals that do not achieve excellent financial performance results will be charged an interest rate premium from their lenders.

Higher costs for capital can sometimes make capital inaccessible to poorly performing hospitals. This may be the financial marketplace's analogy to natural selection. When poorly performing hospitals continue to have little or no access to capital, they perform even more poorly and ultimately fail. Conversely, better-performing hospitals always have access to capital, even in challenging and highly competitive situations. Their focus on strategic cost reduction allows them to receive improved access to capital and lower interest rates.

Hospitals with the lowest cost structures reach the highest levels of profitability. Their ability to produce higher operating margins leads to more favorable bond ratings and lower cost of capital. By deploying capital in the most efficient manner, high-performing hospitals are better able to meet the challenges of their environment. Once hospitals achieve excellent strategic cost reduction results, they are able to use favorable performance trends to gain access to capital at the most favorable rates. Ultimately, this favorable equation fuels growth for the hospital, even in the most challenging reimbursement environments.

MANAGEMENT'S ROLE IN STRATEGIC COST REDUCTION

The hospital's CEO sets the tone for achieving strategic cost reduction success. If the CEO values strategic cost reduction as part of the hospital's leadership culture, the hospital will follow. If the CEO pays little or no attention to strategic cost reduction, unfavorable results will inevitably follow. The CEO imprints his value system on senior management, making the managers also extremely important. If the CEO and senior management together value strategic cost reduction, they will lead the hospital in a manner that achieves excellent results.

Middle management is also incredibly important in sustaining a leadership culture that achieves strategic cost reduction success. Middle managers actually spend the hospital's green dollars. They oversee the day-to-day purchasing of supplies and services, and they supervise the employees who deliver the hospital's services. If middle management is dedicated to the leadership culture of strategic cost reduction, excellent results will follow. If they are indifferent, no amount of accounting or financial control systems will yield strategic cost reduction success. Employees take their cues directly from middle managers. Employees who are encouraged to use their time, supplies, and purchased services effectively will achieve excellent strategic cost reduction results.

THE BOARD'S ROLE IN STRATEGIC COST REDUCTION

The board of directors at the hospital plays a critical role in setting the tone for achieving strategic cost reduction success. Ideally, the board emphasizes and supports the CEO's embrace of strategic cost reduction. If it does not, a culture may develop in which certain managers, physicians, employees, or vendors are likely to defeat strategic cost reduction initiatives. If this happens, the hospital cannot thrive, even with the best of intentions of the CEO and senior management. If the board is supportive, however, the CEO and senior management team can guide the hospital to strategic cost reduction success.

THE PHYSICIAN'S ROLE IN STRATEGIC COST REDUCTION

Physician leaders and practicing physicians within the hospital are also critically important when achieving strategic cost reduction success. Physician leaders who work collaboratively with members of senior and middle management set the tone for achieving ideal results. Physician leaders who are combative or confrontational are not likely to facilitate a culture that is conducive to strategic cost reduction success. In other words, physicians, like middle management and employees, can make or break a strategic cost reduction culture. If physicians help hospitals maximize labor productivity and use supplies and services that produce the best possible results for patients at the lowest possible cost, then excellent strategic cost reduction results will follow.

Collaboration in the vascular suite

The senior management of a community hospital initiated collaboration discussions with its three vascular surgeons to analyze all costs associated with invasive studies performed in the hospital's peripheral vascular laboratory. A team of technicians, nurses, physicians, and two department directors from radiology and vascular services evaluated different approaches to prestudy care, supplies consumed during vascular studies, and poststudy care. The team evaluated alternatives and created protocols that met the needs of about 80 percent of all vascular patients. These protocols specified standard approaches to patient care and supplies used during the vascular studies. In total, the team was able to reduce the labor and supply costs associated with invasive vascular studies by 25 percent. An added positive benefit was reducing prestudy and poststudy patient length of stay by 25 percent. Collaboration at this hospital led to a true reduction of $80,000 green dollars annually as well as a better-quality patient experience because of a shortened hospital length of stay.

APPROACHES TO ACHIEVING THE BEST RESULTS

Achieving excellent strategic cost reduction success is a leadership challenge. Developing a hospital culture that values strategic cost reduction begins with a committed CEO and senior management. This group has the ability to engage all of the various hospital constituencies necessary to attain excellent results. Achieving strategic cost reduction success requires the creation of a culture that listens to patients, middle management, physicians, and employees and then values all that is said. A leadership team that knows the value of green dollars and knows that the team itself must set the best example for the organization ultimately empowers the hospital to succeed.

IS YOUR HOSPITAL READY?

The following questions can help determine whether your hospital is ready for strategic cost reduction:

1. Does your hospital have a current strategic plan with short-term and long-term goals?
2. Does your hospital have a strategic financial plan with short-term and long-term goals and a capital structure plan to support those goals?
3. Is your hospital's CEO and senior management team committed to achieving strategic cost reduction success?
4. Are your hospital's board members knowledgeable about strategic cost reduction and their role in supporting cost reduction success?
5. Do your hospital's middle management, physicians, and employees support the process of evaluating all hospital costs and reducing them to the lowest possible levels without compromising clinical quality or customer service?

Affirmative answers to these questions means your hospital is ready to proceed with strategic cost reduction. The best approach to achieving strategic cost reduction success is to create a leadership culture that values the lowest possible cost structure while simultaneously providing the best possible care for patients. These results cannot be achieved by delegating responsibility to outside consultants. The pathways to achieving strategic cost reduction success start with the commitment of the CEO and senior management. What it takes to succeed is the subject of Chapter 2.

CHAPTER 2

What Does It Take to Succeed?

STRATEGIC COST REDUCTION success takes a leadership philosophy and culture that enable a hospital to achieve the lowest possible costs without compromising quality patient care and excellent customer service. Successful strategic cost reduction involves a two-step process. The first step is to reduce green-dollar costs to the lowest level consistent with the hospital's quality and service standards. This means that labor costs coincide with high employee productivity and that supplies and services are purchased in the most efficient manner. This also means clinical utilization costs reflect the clinical goals of no overutilization or underutilization. The second step to successful strategic cost reduction is to maintain the lowered green-dollar costs at levels necessary to achieve long-term strategic goals.

COMPELLING CASE FOR COST REDUCTION

Successful strategic cost reduction begins with the hospital's leadership making a compelling case. The case may be achieving a strategically important initiative, such as building a new patient tower to replace outdated inpatient facilities. Alternatively, the compelling case may be lowering costs to improve the hospital's competitive position so that it may win market share in a highly competitive community. Another compelling case might be permanently turning around a troubled hospital's deteriorating financial situation so that it may avoid closure or sale to a for-profit hospital chain. Successful fund-raising always begins with developing an engaging case that motivates potential donors to contribute financially to the cause. For example, a case might be to build a new cancer center to lower the rate of cancer deaths in a community. Another case could be to raise funds to build a new emergency department to provide care more promptly to local emergency patients. Hospital leaders have many, varied opportunities to create a compelling case for strategic cost reduction. All cases must come from a compelling strategic goal to be accomplished in the future.

Hospitals facing a financial crisis may, in a moment of panic, contemplate a weak case for cost reduction. One example of a weak case is, after years of operating in the red, cutting costs temporarily to break even financially. Another weak case is improving financial performance to achieve an incentive compensation goal so that senior management can receive their year-end bonuses. Yet another weak case is improving the short-term performance of a hospital merely to avoid triggering a debt covenant or to ensure that the hospital's year-end financial audit will deem the organization an ongoing concern.

> **Bold vision of the future**
>
> The CEO of the smallest hospital in a three-hospital community had a bold vision of the future for healthcare. The largest hospital was a successful investor-owned, for-profit organization, and the remaining two hospitals, the medium-sized and smallest, were community, not-for-profit organizations. The CEO's vision was to improve the hospital's financial performance to exceed that of the other two hospitals in the community and then to institute

a merger with the other not-for-profit hospital. The CEO's vision was to create a new not-for-profit health system that would rival the scope of services at the large, investor-owned hospital.

The CEO began executing this vision with a comprehensive strategic cost reduction initiative at the hospital. Over a two-year period the hospital succeeded in becoming the strongest financially in the community. In the third year, the merger, which the CEO conceived earlier, was consummated. In the fifth year, the newly combined health system surpassed the investor-owned hospital in size and scope of clinical services. Eventually, the new health system expanded to become one of the most successful regional health systems in the state.

This strategic success story began with a bold vision of the future and an effective use of strategic cost reduction.

Fundamentally, success with strategic cost reduction should always begin with the hospital's CEO articulating a compelling case for the cost reduction. Once the case has been conveyed, the CEO's personal dedication becomes critically important.

COMMITTED CEO AND BOARD OF DIRECTORS

The hospital's CEO is absolutely the best person to articulate the case for strategic cost reduction. Only the CEO can mobilize the entire hospital and all of its constituencies to understand and adopt the appropriate case for strategic cost reduction. Only the CEO has the constituencies' trust and confidence to the extent needed to motivate them to make the changes necessary to achieve strategic cost reduction goals. The CEO can inspire the entire hospital's creative energy and assure naysayers that all aspects of the hospital's cost structure will be critically examined and that no "sacred cows" will inhibit the efforts in reducing costs.

It is important that the CEO also stress that the hospital's board of directors is firmly behind strategic cost reduction endeavors. Reassuring the hospital's constituencies, such as middle management, physicians, and employees, that the board firmly supports strategic cost reduction helps ensure that politics will not be a factor while the hospital critically examines itself and makes changes crucial to achieving strategic cost reduction goals.

The CEO also needs to act as the hospital's chief communication officer during strategic cost reduction efforts. The CEO must invite and inspire hospital constituencies to put forth their best efforts to examine green-dollar costs, reduce them without compromising quality, and implement the fundamental cultural changes that are critical in sustaining strategic cost reduction initiatives in the future. No matter how gifted, however, the CEO will need the strong support of senior management to introduce and implement strategic cost reduction successfully.

COHESIVE SENIOR MANAGEMENT

Full commitment and energetic support for strategic cost reduction by the hospital's senior management, such as the chief operating officer (COO) and vice presidents, is absolutely essential to achieve success. Lack of support or passive-aggressive behavior among any member of the senior management team can derail a strategic cost reduction initiative, even if it has the full support of the hospital's CEO. The chief financial officer (CFO) is critically important as well. Showing a complete dedication to strategic cost reduction allows the CFO to enhance the probability of success. On the other hand, CFOs can undermine strategic cost reduction efforts if they do not show the utmost respect for the efforts of operating vice presidents and middle management.

A cohesive and energized senior management team sets the positive example needed for strategic cost reduction. They need to show strong support for the CEO and strong support for middle management if the initiative is to succeed. The senior management team can be thought of as the orchestra and the CEO as its conductor. A sour note among the orchestra will resound throughout the hospital and make it impossible for strategic cost reduction to prevail. On the other hand, a well-coordinated and highly motivated senior management team provides the foundation for success and sets the stage for middle management to be fully engaged in strategic cost reduction.

ENERGIZED MIDDLE MANAGEMENT

Perhaps no other group is collectively more critical to strategic cost reduction success than the hospital's middle management. They are the principal figures in every hospital because they oversee where green dollars are actually spent. They hire and fire employees. They purchase the supplies and the services needed to deliver patient care day in and day out. They know where the money is spent, and they know where costs can be reduced without compromising quality or service. Therefore, they are profoundly important to strategic cost reduction success.

It is highly critical for middle management to favorably influence the perceptions of physicians and employees. In successful strategic cost reductions, middle management positively engages the interest of physicians and employees. A skilled and resourceful middle management team can extract creative cost reduction ideas from committed physicians and employees and implement them. Although it may be impractical for senior management to approach each and every physician and employee individually about cost reduction initiatives, it is well within the grasp of middle management to engage each and every physician and employee in their respective departments to contribute ideas for achieving cost reduction goals. Indeed, middle management plays an extraordinarily prominent role in engaging the support of physicians and employees during the strategic cost reduction process.

MOTIVATED PHYSICIANS

Physicians who understand the compelling case for strategic cost reduction can be highly motivated to contribute to its success. If, for example, the case for strategic cost reduction focuses on building a new patient tower that will provide state-of-the-art technology, efficiently designed patient care space, and more opportunities for physicians to take the absolute best care of their patients, the physicians will be enthused about and dedicated to helping achieve the cost reductions necessary to make the new patient tower a reality. It has been said that the most expensive technology in any hospital is a ballpoint pen in the hands of a physician, but highly motivated physicians can wield that pen in a fashion that contributes to strategically reducing costs.

In successful strategic cost reduction initiatives, physicians are able to critically examine utilization costs for clinical diagnostic and therapeutic procedures. They are then able to articulate and support changes in those clinical practices that can reduce costs while preserving quality. The physician's examination of alternatives for supplies, pharmaceuticals, and services used for patient care can also yield profoundly constructive ideas for strategic cost reduction. Physicians can also have a great influence on the number, skill levels, and deployment of hospital employees. Physicians who are open to considering alternatives can be powerful allies for senior management to reach strategic goals in labor cost reduction. Clearly, a highly enthusiastic and cooperative medical staff contributes a great deal to achieving success in strategic cost reduction.

ENGAGED EMPLOYEES

Employees exert a profound influence on hospital costs. Labor costs, supply costs, service costs, and clinical utilization costs are all directly influenced by employees. Highly productive employees enable hospitals to achieve the lowest labor costs. Actively engaged employees can influence the costs of supplies and services purchased on behalf of patients. Even the simple act of having employees turn off the lights in unoccupied rooms can affect hospital utility costs.

To paraphrase the late Senator Everett Dirksen (R-Ill.), "A dollar here, a dollar there, pretty soon it adds up to real money." The same is true with hospital costs. Employees can represent the difference between success and failure in achieving strategic cost reduction goals.

MEASURED USE OF CONSULTANTS

The presence or absence of consultants can affect the outcome of strategic cost reduction. Some hospital CEOs and senior management teams believe they can delegate cost reduction efforts to consultants while standing at the sidelines or pursuing other priorities. Others mistakenly believe they need no help in cost reduction efforts and never consider the use of consultants. The measured use of consultants can enhance a hospital's success with strategic cost reduction. However, consultants should never lead a strategic cost reduction initiative. Consultants cannot substitute for a hospital's CEO in creating a compelling case for strategic cost reduction and communicating that case to the hospital's constituencies. Consultants alone cannot develop practical ideas for cost reduction and then ensure successful execution of those ideas.

Ultimately, the thorough implementation of changes necessary to achieve strategic cost reduction success will always rest with the hospital's CEO, senior and middle management, physicians, and employees. Consultants can act as effective facilitators—perhaps that is their best and truest contribution to a hospital contemplating strategic cost reduction. If consultants can facilitate the idea-generating process and, through their diverse professional experiences, assure a hospital that the ideas can be successfully implemented, then consultants can surely be positive contributors to the strategic cost reduction process.

IMPLEMENTATION: THE KEY TO LONG-TERM SUCCESS

A compelling case; a committed CEO and senior management team; motivated middle management, physicians, and employees; and the measured use of consultants are all crucial aspects of strategic cost reduction success. What ultimately matters, however, is that the hospital has the ability to implement strategic cost reduction initiatives and sustain them over time. The first step is generating a number of cost reduction initiatives that can be implemented without compromising quality or service. Once the initiatives are identified and validated and the hospital is prepared to move forward, implementation is paramount. Participation of the CEO and senior management staff is vital when it comes to implementation; they must prepare the hospital for change. Change does not only relate to specific initiatives; it also relates to the culture of the hospital that is so important to achieving long-term strategic cost reduction. The compelling case for strategic cost reduction should be prominently discussed during and after implementation. After all, it is this case that initially compelled the hospital to achieve strategic cost reduction success.

Another important aspect of implementation is to predetermine the effect of all strategic cost reduction initiatives. Staff changes, supply and service changes, and process changes must all be carefully evaluated before the moment of implementation. After implementation, it is important that the hospital is prepared to fine-tune changes as necessary to ensure that all initiatives are fulfilling their cost reduction goals without compromising patient-care quality and customer service. Next, it is absolutely essential that all changes be frequently audited during the first six months to ensure that the intended green-dollar effect is achieved.

It is highly important for the CEO and senior management team to provide middle management, physicians, and employees with additional education and leadership-development opportunities after implementing strategic cost reduction. Strategic cost reduction involves not only making technical changes but also transforming leadership effectiveness. As part of this transformation, it is important that management at all levels of the hospital improve leadership and communication skills. Success in sustaining strategic cost reduction initiatives into the future depends, in part, on instilling changes in the hospital's budgeting process. The thought process that generated strategic cost reduction initiatives must be fully integrated into the hospital's budgeting cycle to ensure that changes are maintained and enhanced in the future.

DOES YOUR HOSPITAL HAVE WHAT IT TAKES TO SUCCEED?

The following questions can help determine whether your hospital has the necessary strengths to succeed with strategic cost reduction:

1. Can your hospital's CEO make a compelling case for strategic cost reduction?
2. Is your hospital's CEO personally committed to pursuing strategic cost reduction?
3. Do senior and middle management teams support the strategic cost reduction process?
4. Are your hospital's physicians and employees willing to participate in developing strategic cost reduction initiatives?
5. Is your hospital committed to implementing and sustaining strategic cost reduction to achieve long-term strategic goals?

Affirmative answers to all five questions are essential to ensure that your hospital has the appropriate foundation to proceed with strategic cost reduction. Nevertheless, in spite of their best efforts, some hospitals fail in their quest to achieve strategic cost reduction success. Chapter 3 explores some of the most common reasons for failure.

CHAPTER 3

Why Do Some Hospitals Fail?

MANY HOSPITALS FAIL to lower costs strategically. That is one of the reasons approximately one-quarter of all hospitals operate in the red and must rely on savings or loans to meet their everyday financial obligations. What are some of the common reasons for this failure? One fundamental reason is that cost reduction is very challenging; it takes the CEO and senior management team's supreme commitment to achieve excellent financial results. Not every senior management team has the discipline and dedication it takes. Not every senior management team understands the strategic imperative to reduce costs. Without a strategic plan, strategic cost reduction will fail.

Another common reason for failure is that some hospital leaders view cost reduction as a project that can be crossed off on a to-do list or a once-a-year exercise during budget time. Although a project or budget-cutting approach to cost reduction might produce favorable short-term results, this approach rarely produces successful long-term results. Sustainable strategic cost reduction takes the steadfast commitment of the CEO and the senior management team.

Failure also occurs when strategic cost reduction is delegated to consultants. Some hospital CEOs and senior management teams mistakenly believe they can hire outsiders to do a cost reduction project. Again, this may yield short-term results, but it does not guarantee long-term success. The obsessive use of benchmarks to guide cost reduction efforts is a variation of the consulting approach. Although benchmarks such as productivity ratios, profitability rates, and bond ratings are excellent management guides, benchmarking alone rarely leads to long-term success in strategic cost reduction.

Sometimes strategic cost reduction fails simply because of poor execution of individual cost reduction initiatives. It is not enough for hospital managers to identify cost reduction initiatives. To truly reduce costs, those initiatives must be successfully implemented and sustained. The ability to implement is a highly important element on the pathway to long-term success in strategic cost reduction. The failure to acknowledge the challenges of strategic cost reduction, an uncommitted leadership team, the delegation of an entire cost reduction process to a consultant, and poor execution of initiatives are all factors that can contribute to a failed strategic cost reduction effort.

FAILURE OF LEADERSHIP

Some strategic cost reduction efforts fail because the CEO is disengaged or does not see the value in personally leading strategic cost reduction. It has become customary for some CEOs to be seen as "outside" oriented, responsible for everything but the day-to-day running of the hospital. A leader with this outside orientation is a CEO who rarely sees the value in personally leading a strategic cost reduction initiative. It is not uncommon for these outside-oriented CEOs to delegate cost reduction to senior management, the finance staff, or consultants. Although the finance staff has the technical knowledge to understand hospital costs, they have neither the leadership authority nor the hospital-wide influence to initiate and maintain true strategic cost reduction effectiveness. The absence of CEO commitment will also inevitably lead the hospital's middle management and physicians to conclude that strategic cost reduction is not crucially necessary. Without full support from these two constituencies, strategic cost reduction will fail.

NO STRATEGIC PLAN FOUNDATION

Strategic cost reduction needs a strategic plan for its foundation. Some hospitals do not have a grasp on this fundamental point: They fail in strategic cost reduction because they failed to create a strategic plan first. The strategic plan provides the foundation for making the compelling case for strategic cost reduction. It is true that hospitals can cut costs in the absence of a strategic plan, but it is also true that they cannot cut costs strategically, and therefore efficiently, without a strategic plan.

> **Strategy vacuum**
>
> For decades, a community hospital in a competitive two-hospital town had enjoyed parity with another hospital in clinical services, facilities, prestige, reputation, and market. That parity evaporated when the other hospital hired an aggressive CEO who created a grand strategic vision of the future and then proceeded to execute the vision with great vigor and success. Over a ten-year period, market share moved from 50/50 to 80/20.
>
> The losing hospital resisted all efforts to create its own strategic vision and plan for the future. Its CEO pursued aggressive cost reduction measures every year, yet never cut enough costs to stem the mounting financial losses. Physicians and employees became demoralized with this constant cost cutting and abandoned the losing hospital in droves, taking with them most of the previously loyal patients.
>
> Continuous cutting of employees, supplies, and programs with no establishment of strategic direction finally cost the CEO her job. By the time she left office, the hospital was nearly bankrupt

and susceptible to closure. This CEO's failure to create effective strategic planning nearly cost the community its second hospital. This sad story is a compelling confirmation that cost cutting without a strategic plan foundation is a course of action doomed to failure.

LACK OF TRUST

Mutual trust among all levels of the hospital is required for successful strategic cost reduction. Middle management, physicians, and employees must trust the CEO when he makes the case for strategic cost reduction. If they do not trust the case, they will not accept it and certainly will not work toward its successful achievement. On the other hand, the CEO and senior management must trust middle management, physicians, and employees to achieve success in strategic cost reduction. The CEO and senior management must trust that those closest to the spending of green dollars will know how best to reduce costs without compromising quality or service. If the CEO and senior management team do not have trust, they will be prone to overusing consultants and external benchmarks in their attempt to reduce hospital costs. These efforts are destined to fail.

OBSESSION WITH BENCHMARKS

Benchmarks are, by their very nature, conglomerations of averages. Benchmarks alone can never guarantee a success in strategic cost reduction. They are an information tool, not a management tool. Merely presenting a benchmark, regardless of how credible to a department director or physician, will not motivate that person to make sustained changes to lower costs. Benchmarks are very useful as directional guides, however. If a labor-productivity benchmark suggests that a department is significantly overstaffed, the benchmark may lead to a thorough evaluation of the department's operation to determine where staffing improvements can be made. The benchmark will often be subject to a thorough evaluation as well; a physician leader or department director will rarely accept a performance benchmark without challenging that benchmark's relevance. It has been my experience that department directors spend more time arguing about a benchmark's relevance if the benchmark is presented as the sole rationale for reducing costs than they do talking about how to improve the department's operation.

OVERDEPENDENCE ON CONSULTANTS

Consultants give advice; they do not lead. An army of consultants cannot accomplish what one highly motivated CEO can achieve when it comes to strategic cost reduction. Consultants can never achieve sustained strategic cost reduction excellence. Consultants cannot change a hospital's culture to make it more conducive to strategic cost reduction efforts. At best, they can assist a highly motivated and capable senior and middle management team, but they cannot substitute for that team.

Hospitals have wasted many millions of dollars on consultants in the quest to achieve cost reduction results. When a hospital's leadership team depends on consultants to reduce costs, they are apt to be labeled as hypocrites by physicians and employees because oftentimes the consultants' fee costs hundreds of thousands, or even millions, of green dollars. In other words, excessive use of consultants sometimes leads to extreme resistance and resentment toward strategic cost reduction.

Consultants, used judiciously, can help a hospital's leadership team create a culture that values and practices strategic cost reduction. Consultants alone cannot produce excellent results. A hospital leadership team that believes otherwise is vulnerable to failure.

POOR EXECUTION SKILLS

Another reason for failing to achieve excellent strategic cost reduction results is the inability to execute and implement cost reduction ideas. In some cases, a hospital successfully identifies initiatives and strategies to reduce labor, fringe benefits, supply, and purchased service costs. They falter, however, when implementing those initiatives. Sometimes this is a result of senior management's lack of commitment. This occurs when sacred cows get in the way of implementation; they thwart otherwise sound initiatives because they try to impose on someone powerful enough to resist initiatives. In this sense, failure is a result of not holding senior or middle managers accountable for achieving strategic cost reduction success.

At other times, the genesis of failure is not monitoring implementation of cost reduction initiatives closely enough to ensure that they achieve their desired outcomes. Failure to monitor can lead to stealth cost increases, which can destroy the desired results of strategic cost reductions. For example, reduction in labor costs associated with not replacing a retired employee poses minor long-term effects if corresponding increases are made in overtime, agency staff, and additional hours worked by other employees who fill in for the paid hours saved when the employee retired. Therefore, poor execution skills often result in a strategic cost reduction failure.

PROJECT MENTALITY

Sometimes failure to succeed in strategic cost reduction results from not understanding the importance of the hospital's culture. Short-term cost reduction success, if the effort is viewed as a "project," is destined to atrophy over time. Hospitals that are successful with strategic cost reduction understand that they must permanently change the hospital's culture to sustain cost reduction results. Indeed, concentrated focus on cost reduction for a short time may lead to positive short-term results. However, these results will atrophy if the hospital's leadership culture does not sustain the required focus and attention on strategic cost reduction.

When cost reduction is viewed as a project or campaign, usually results are initially positive, but the results will be replaced with long-term failure. Nothing is worse for a hospital's leadership team and employees than to go through a cost reduction roller coaster year after year at budget time because they fail to achieve long-term and sustainable cost reduction results.

FINGER-POINTING

Finger-pointing during strategic cost reduction efforts is another reason for failure. It is also a convenient excuse for failure. CEOs point the finger at senior management, senior management points the finger at department directors, and department directors point the finger at physicians or employees for failing to produce desired results.

When internal sources are exhausted, hospital CEOs and senior management can then resort to blaming consultants for failing to deliver success. The same consultants who were heralded as experts are often labeled expensive failures if strategic cost reduction results are not in line with expectations. If consultants cannot shoulder the entire blame for failure, a hospital can then turn its attention to blaming payers, competitors, and the ever-present villain, the government. Hospital leaders who are prone to finger-pointing can ultimately find the true source of strategic cost reduction failure by using a very inexpensive piece of technology: a mirror.

LACK OF RESOLVE

When strategic cost reduction success is not accomplished, the real explanation ultimately leads to the door of leadership, where the responsibility belongs. Achieving and sustaining strategic cost reduction success is difficult. It takes leadership's resolve. If that resolve is not present, no amount of benchmarking, consulting assistance, or short-term cost-cutting programs will lead to strategic cost reduction success. A hospital that achieves and sustains strategic cost reduction excellence does so with the full and complete engagement of its CEO and senior management team. Anything less than the full resolve of these important leaders will undoubtedly end in failure.

CAN YOUR HOSPITAL AVOID FAILURE?

The following questions can help your hospital decide if it is potentially subject to failure with strategic cost reduction initiatives:

1. Do your hospital's middle management, physicians, and employees lack trust in the CEO and senior management?
2. Are your hospital's leaders overly enamored with benchmarks?
3. Are consultants usually the first response to solving challenges at your hospital?
4. Do your hospital's CEO and senior management believe strategic plans are unnecessary during this period of rapid healthcare-industry change?
5. Does your hospital have a history of failing to sustain long-term initiatives?

An affirmative answer to any of these questions should cause your hospital to pause before beginning a strategic cost reduction initiative. Although there are many opportunities for failure with strategic cost reduction, successful hospital leadership teams can avoid or overcome them all. In doing so, they set a great example of leadership effectiveness. An extraordinary leadership success story is the subject of Chapter 4.

CHAPTER 4

Extraordinary Success Story at Northeast Georgia Medical Center

NORTHEAST GEORGIA MEDICAL CENTER faced a series of strategic and financial dilemmas. While running at full capacity, the hospital's operating margin was approximately break-even. Capacity constraints prevented the hospital from increasing revenue through volume growth. The inpatient beds were full, the operating rooms were constantly booked, the emergency department was overflowing, and outpatient services were strained to the limit. Staff productivity levels were relatively low, and the average age of plant and equipment was relatively high because of an aging facility. Growing indigent care problems combined with a relatively low Georgia Medicaid payment rate provided immense challenges for this full-service regional hospital with more than $455 million in annual revenue.

    A new CEO, James Gardner, assumed leadership of Northeast Georgia Medical Center when the previous president retired after more than 40 years of distinguished service. Management and employees were comfortable with the hospital's performance. The new CEO was not. It quickly became apparent to him that the hospital's capacity problems, aging physical plant and medical technology, and relatively complacent management were potential impediments for future change. Although the hospital was stable financially, it was not thriving. Although the physical plant and medical technology were functional, they were not state-of-the-art. Although the medical staff was diverse and quality oriented, it was not engaged in meaningful collaboration with the hospital leadership. The hospital's capital plan, which included replacing a major portion of the inpatient facility, proved to be unrealistic; current financial performance would not allow the hospital to access sufficient capital to execute the facility and technology improvements needed to make the hospital state-of-the-art again.

As CEO, Gardner viewed these challenges as windows of opportunity to change. He began by mobilizing Northeast Georgia Medical Center to allow what he characterized as a "strategic turnaround." To do this, he engaged the board of directors, senior management, middle management, medical staff, and employees in the dual tasks of completing a comprehensive strategic plan for the future while strategically lowering the hospital's cost structure. The board of directors insisted that these dual goals be pursued while simultaneously improving customer service, clinical quality, and employee morale. Although these may seem like contradictory goals, Northeast Georgia Medical Center proved they could be accomplished. The medical center's success story is a road map for all hospitals, regardless of size, for how to successfully implement strategic cost reduction.

MAKING THE CASE

Northeast Georgia Medical Center's first challenge was what its new CEO described as "making the case." When financial collapse is eminent, making the case for change becomes obvious: Either performance improves, or the hospital perishes. But what happens when a hospital is seemingly satisfied with its less-than-stellar performance? That is the question Gardner faced first. He and his senior management team analyzed recent historical financial and performance trends. Next, they ordered a careful review and comprehensive update of the hospital's future capital needs in conjunction with commissioning a new strategic plan. This review demonstrated that immediate capital needs were $150 million, not $80 million as previously thought. Long-term capital needs of $450 million were identified for the next eight to ten years.

Strategically lowering the hospital's cost structure became a critical priority for gaining access to capital so that strategic initiatives could be founded; this is now clearly articulated as part of the strategic planning process. After understanding strategic capital needs of the future, Gardner and his team then faced the decision of whether or not consultants were needed to help the hospital achieve strategic cost reduction goals. Gardner considered doing it alone, hiring a large consulting firm, or using a small consulting firm with a "facilitated" approach to strategic cost reduction. He chose the small-firm approach and retained a highly experienced adviser to help his leadership team accomplish strategic cost reduction.

DISTINGUISHING FEATURES OF THE STRATEGIC COST REDUCTION APPROACH

Gardner, his strategic cost reduction adviser, and the hospital's senior management team devised a unique and extraordinarily successful approach to strategic cost reduction. The approach began with the recognition that external benchmarks would be of only limited value. Previous internal benchmarks during years of exceptionally good financial performance were used as an alternative to external benchmarks. It was agreed at the outset that the adviser would facilitate the strategic cost reduction process and senior management would lead. Gardner and his senior management team set five goals with help from the strategic cost reduction adviser:

1. Reduce annual operating expenses by $20 million per year without compromising quality.
2. Improve physician and employee morale.
3. Improve customer service throughout the hospital.

4. Create an opportunity for meaningful physician and employee input into the strategic cost reduction process.
5. Implement the cost reduction initiatives to position the hospital for a successful bond issue one or two years hence.

TASK FORCE METHODOLOGY

The hospital's adviser recommended organizing senior and middle management, approximately 100 leaders, into task forces for identifying strategic cost reduction initiatives. Nine task forces of approximately 11 people each were organized. Middle management, including directors and managers, constituted eight of the task forces. Senior management, vice presidents, and Gardner's direct reports made up the ninth task force. The eight middle-management task forces were organized so that membership was highly diversified. Clinical and nonclinical directors and managers were intermingled. Long-tenured and recently hired directors and managers were also intermingled. This diversification meant that the task forces included directors and managers who barely knew each other at the beginning of the strategic cost reduction process.

Once the task forces were organized, the hospital's adviser conducted training sessions and developed ground-rule guidelines for how the task forces were to operate. A detailed schedule was developed, encompassing approximately 60 days of intensive work by task force members.

TASK FORCE KICKOFF RETREAT

The strategic cost reduction initiative was formally launched at Northeast Georgia Medical Center with a kickoff retreat for all members of senior and middle management. The retreat was used as a training session conducted by Gardner, senior management, and the hospital's adviser. Here, goals were assigned to each task force and the hospital's adviser explained the concept of green dollars in great detail. He explained that cost reduction initiatives could not be counted toward the $20 million goal unless the hospital stopped "writing checks." Hypothetical, theoretical, or hoped-for savings could not be counted toward the strategic cost reduction goal. Only true green-dollar reductions in salaries, fringe benefits, supplies, or purchased services could be counted. Checks for $20 million a year would have to cease for this aggressive strategic cost reduction goal to be achieved.

Immediately after the kickoff retreat, Gardner conducted what he described as "town meetings" with all physicians and employees. This required a series of exhausting, around-the-clock meetings so that the medical staff of 425 and employee staff of 3,500 could all receive personal communication from the CEO about why strategic cost reduction was needed and a detailed explanation of the process. These town meetings, held at the beginning, middle, and end of the strategic cost reduction process, garnered high praise from physicians and employees.

Gardner and senior management went to great lengths to communicate the cost reduction process to the board of directors. The message to the board was that the financial performance needed to improve by $20 million annually and that the performance of the hospital's leadership needed to improve. Additionally, Gardner made the commitment to simultaneously boost employee morale and customer service while achieving strategic cost reduction goals. Although members of the board appreciated the aggressive approach, many were skeptical that all of Gardner's ambitious goals could be achieved. Time proved the skeptics to be entirely wrong.

TASK FORCES AT WORK

For 60 days, the hospital's nine task forces met approximately three times per week. Other major hospital priorities were temporarily suspended or curtailed to give all members of management time to focus intimately on strategic cost reduction. They brainstormed with each other and challenged each other. Each task force had a specific green-dollar goal to achieve. Goals were assigned to each task force proportionate to the total annual-expense dollars controlled by task force members. In other words, if the 11 task force members in total controlled 10 percent of the hospital's operating costs, then the task force was assigned 10 percent of the $20 million overall goal, or $2 million. At first, these brainstorming sessions were a learning experience for task force members. The members quickly realized, however, that the hospital's CEO had entrusted them with a seemingly impossible cost reduction challenge. Gardner's trust provided an extraordinary level of inspiration for directors and managers to achieve their assigned task force goals.

The hospital's adviser met with each task force approximately every 10 days throughout the 60-day brainstorming period. His role was one of teaching; facilitating; and, occasionally, administering sharp critique. The adviser used his previous experiences with strategic cost reduction to help directors and managers focus on realistic ways to reduce costs while preserving patient care quality and customer service. He challenged them to restructure staffing patterns around employee attrition. He cajoled them into considering supply alternatives and different approaches to purchased services that could accomplish the needed tasks to lower costs. Nothing went unnoticed; every hospital cost was examined in great detail by the nine task forces.

The adviser worked particularly closely with the ninth task force, which comprised Gardner and his senior management team. This task force had a very aggressive goal for green-dollar cost reduction, and each vice president, and Gardner himself, was expected to put forth ideas to make sufficient cost reductions to reach their assigned goal. This ninth task force carefully evaluated management layers within the hospital, staffing, fringe benefits, and purchased services for a wide variety of hospital needs. They also considered renegotiating major supply contracts and making clinical program changes to achieve their goal.

On Friday afternoon, each task force was required to submit directly to Gardner a weekly progress report, detailing initial cost reduction ideas and works in progress. Weekly progress reports were reviewed by Gardner and senior management on Monday morning. Senior management used these weekly reports to help the task forces evaluate the cost reduction implications of their ideas. In this way, senior and middle management were working together to achieve the common goal of $20 million in green-dollar cost reductions. Throughout this process, each member of every task force was expected to meet with their department's physicians and employees. In this way, the task force process gained access to the ideas of all physicians and employees throughout the hospital. This grassroots approach not only created a sense of widespread teamwork throughout the hospital, but it also produced some extraordinarily creative ideas for reducing costs.

CHALLENGES ENCOUNTERED BY TASK FORCES

During the 60 days of intensive task force work at Northeast Georgia Medical Center, several challenges were encountered. It quickly became apparent that some directors and managers were highly comfortable with the status quo and would resist change with great vigor. The hospital's culture had not previously been one of high accountability for members of senior and middle management. Some medical staff constituencies also had a strong resistance to change. The paternalistic approach to leadership of the previous CEO had permeated the hospital's management ranks.

Gardner met these challenges in a straightforward manner. His strong commitment to a grassroots approach to strategic cost reduction reinforced a higher level of accountability as part of the hospital's new leadership culture. He made it clear that no one was exempt from this accountability, including himself and senior management. During this process, sacred cows (of which there were several) were exposed and put out to pasture. Gardner emphasized that he expected open and honest communication and a positive attitude about the strategic cost reduction process from everyone. At the end of the 60-day period, all task forces presented initiatives that achieved their green-dollar goals. Collectively, a total of $25 million in cost reduction initiatives were identified and ultimately the hospital's senior management approved $24 million in changes. In addition to cost reduction proposals, each task force recommended employee morale and customer service improvement's that were considered and approved by senior management for immediate implementation.

LEADERSHIP REVIEW

While all of the other task forces were hard at work identifying strategic cost reduction initiatives with physicians and employees, Gardner and his senior management team of the ninth task force were engaged in a comprehensive evaluation of the hospital's organizational structure and an individual evaluation of all management personnel. Their primary objective was to flatten the organization by reducing layers of management wherever possible. Their secondary objective was to identify members of the management team for future promotion and identify those individuals whose skills no longer matched the hospital's future needs.

Two vice president and two director-level positions were eliminated because of the leadership evaluation. About a dozen middle management positions were combined or consolidated, and several prominent members of management left the hospital because of performance impediments. Once the leadership review was completed and the changes were implemented, Northeast Georgia Medical Center focused heavily on organizational development. The new leadership team has been extremely successful in implementing and sustaining strategic cost reduction initiatives and creating an updated leadership culture that emphasizes accountability.

DECISIONS, DECISIONS, DECISIONS

Northeast Georgia Medical Center's senior management considered more than 300 recommended initiatives for cost reduction that were submitted by the task forces. Senior management developed detailed implementation plans, and a timetable was established for each initiative. Staffing reductions of approximately 200 positions were implemented, mostly through attrition and payroll practice changes; major changes were made in the employee benefit program; and a myriad of supply and service changes were carried out. The appendices described at the end of the book highlight examples of Northeast Georgia Medical Center's strategic cost reduction initiatives and the task force process they used so successfully.

Postimplementation results were dramatic. As planned, the hospital's financial performance immediately improved by approximately $2 million per month after the strategic cost reduction initiatives were implemented. The strategic cost reduction process created a tremendous sense of ownership among managers, directors, physicians, and employees. This successful outcome facilitated a true transformation of Northeast Georgia Medical Center's leadership; it energized senior and middle management and gave physicians and employees a new sense of ownership in their hospital's success.

One year following the strategic cost reduction initiative, Northeast Georgia Medical Center posted the highest operating income and margin in its history. Its stellar financial and leadership performance enabled Northeast Georgia Medical Center to access the bond market for needed capital to complete its strategic plan objectives. Its bonds received the highest possible rating and insurance coverage because of its highly improved financial performance. Gardner credited his board, leadership team, physicians, and employees with these profoundly important milestones in Northeast Georgia Medical Center's history and rewarded all employees with their first-ever financial bonus.

LESSONS LEARNED

Gardner describes this successful outcome as a "leadership transformation, with a highly positive financial performance improvement result." He and his hospital's senior management team accomplished a stunningly successful strategic turnaround. Gardner articulates several key lessons learned:

1. Preparation is the key.
2. Constant communication is required.
3. Get it done and over in four months.
4. The CEO must take ownership of this work.
5. Be bold and choose a good adviser.

Department directors and managers also learned several key lessons from the strategic cost reduction process. They articulated their personal lessons learned as follows:

1. We can challenge each other.
2. We can learn from each other.
3. We can teach each other.
4. Together, we can achieve greatness.
5. History will record that this path to greatness began with a new vision of the future and the successful utilization of the task force approach to strategic cost reduction.

HOW DOES YOUR HOSPITAL COMPARE TO NORTHEAST GEORGIA MEDICAL CENTER?

Hospitals that are contemplating embarking on a strategic cost reduction process should begin by asking themselves these key reality-check questions that are based on the experiences of Northeast Georgia Medical Center:

1. Does your hospital have the commitment to renew the

process of strategic planning before or during the actual strategic cost reduction process to ensure that a profoundly important case for change can be articulated to all hospital constituencies?
2. Does your hospital have the discipline to change its corporate culture even if that culture has years of momentum?
3. Is your hospital willing to eliminate senior and middle management positions to set a leadership example for streamlining?
4. Can your hospital significantly reduce the number of employees and improve productivity?
5. Does your hospital have the strength in middle management members to sustain employee productivity improvements?
6. Can your hospital's leadership team work with physicians to make substantive clinical practice changes in drug use, supply utilization, length of stay, clinical test and service utilization, and so on?
7. Can your hospital implement a performance-based retention-and-reward system for management and employees?
8. Will your hospital's board of directors support senior management if challenged by middle management, physicians, employees, and community members during strategic cost reduction implementation?
9. Can your hospital improve employee morale and customer service while simultaneously implementing strategic cost reduction improvements?
10. Can your hospital afford to do nothing? Is status quo an option?

Northeast Georgia Medical Center is an extraordinary example of a hospital that transformed itself using strategic planning and strategic cost reduction. Not only did the hospital succeed with strategic cost reduction, but it also sustained the financial performance improvements while simultaneously enhancing its physician morale, employee morale, and customer services experiences. It is a remarkable hospital led by a remarkable senior management team and board of directors.

Every hospital is unique. Even so, every hospital, regardless of size, clinical focus, or geographic location has opportunities to succeed with strategic cost reduction. Part II highlights these opportunity areas; it explains the place to begin the search for initiatives that can reduce hospital green-dollar costs without negatively affecting patient care or customer service quality.

PART II

Best Opportunities for Strategic Cost Reductions

ACHIEVING SUCCESS IN strategic cost reduction takes a committed hospital leadership team with a laserlike focus on costs that can be permanently reduced without compromising quality or service. The chapters in Part II highlight six key areas of hospital cost reduction opportunities that are available to able leadership teams. Chapter 5 begins with a discussion of management costs and the challenges leadership teams must address to permanently lower management costs. Chapter 6 discusses labor costs and the necessary commitments to permanently improve productivity and lower labor costs. Chapter 7 highlights supply costs, and Chapter 8 provides insights into service costs and where hospital leadership teams should focus to strategically lower service costs. Chapter 9 reviews utilization costs and the opportunities for collaboration between hospitals and physicians that can permanently lower utilization costs. Finally, Chapter 10 highlights the importance of capital costs on a hospital's strategic cost reduction success.

These chapters are intended to provide insights that hospital leadership teams can use to develop cost reduction activities that best fit their individual hospital situations. There are no generic prescriptions that fit every circumstance of every hospital for reducing green-dollar costs. The value of the following chapters is their ability to help leadership teams ask the applicable questions for their hospital's situation. To answer these questions honestly and thoughtfully is to take the beginning step on the pathway to strategically lowering hospital costs.

CHAPTER 5

Management Costs

IT IS INADVISABLE to contemplate a strategic reduction of overall hospital costs without first critically analyzing costs associated with the hospital's management. A thorough and critical examination of the hospital's management structure, such as the number of layers of management between the CEO and the bedside caregiver, is the first step to analyzing management costs. Examination of management costs should include the challenging subject of competence and the excessive costs associated with underperforming or incompetent management. The hospital's CEO and senior management have an obligation to look to themselves first when embarking on strategic cost reduction for their hospital. To do otherwise could diminish their credibility within the hospital and lower the probability of success with strategically lowering labor, supply, service, clinical utilization, and capital costs.

COST FACTORS IN MANAGEMENT

Hospital management costs can be higher than necessary for three critical reasons. First, some hospitals simply have too many managers. This is the case when a hospital uses three layers of management when one or two will suffice. Second, some hospitals have management competence problems. For instance, when a hospital hires an assistant for a poorly performing department director instead of replacing the director, there is a higher-than-necessary management cost. Third, hospitals can have excessive middle management costs related specifically to ineffective senior management. The responsibility for this problem rests squarely on the shoulders of the CEO. Where should a hospital proactively begin the process of strategic cost reduction? In its own executive suite.

SENIOR MANAGEMENT CHALLENGES

First, a reiteration of the definition of senior management: Senior management is the hospital's CEO and the CEO's direct reports. In small hospitals, senior management may be composed of three or four executives. In large medical centers, it may be a dozen or more executives. The skill and competence of every member of the senior management team must be critically and continuously analyzed by the CEO. Are the right leadership skills present? Is the experience of senior executives consistent with their leadership assignments in the hospital? Are executives retained based on their superb contributions to the hospital, or are they retained as a function of momentum and inertia? Is the hospital's organizational chart based on personalities rather than skills or common sense? These questions must be asked and answered by the CEO when any hospital begins the strategic cost reduction process.

> **Two for one**
>
> The CEO of a large community hospital that was part of a multihospital healthcare system had a problem. The system's corporate office set salary ranges too low for hospital vice presidents. In fact, the range was so low that the CEO experienced turnover

in the vice president of support services position every year or two because the CEO was only able to recruit students recently out of graduate school to work for such a low salary. The disruption to this CEO's leadership team was significant, and the costs of the yearly executive search were excessive.

When the CEO approached the corporate office about the situation, she was told that the salary range could not be changed, and her request for a higher range was denied. They offered, however, to let her hire two vice presidents instead of one so that the integrity of the salary range would not be compromised. This ill-advised approach potentially doubled the costs for the vice president position, when a 20 percent increase in the salary range would have solved the problem effectively. In a not-surprising leadership development at the corporate office, the system's CEO was asked to step down not long thereafter.

Additional questions for senior management concern using in-house expertise versus outsourcing expertise. Are full-time senior executives used when functions could be done more cost effectively through outsourcing? Is the hospital carrying the cost of staff, such as secretaries and administrative assistants, to support executive positions that are no longer relevant? Is the hospital overpaying or underpaying for its executive suite talent? It is the job of the hospital's CEO to ask and answer these important questions. A hospital that succeeds with strategic cost reduction will have first succeeded with ensuring that its management costs are as low as possible and consistent with delivering excellent leadership results.

Senior vice president trap

The CEO of a large medical center decided to add a senior vice president of legal affairs position to the senior executive staff. His rationale was that this new position would lower legal costs by reducing dependence on outside law firms, a sound idea in theory.

As an initial step in the strategic cost reduction process, the CEO carefully evaluated the impact on legal expenses of the new senior vice president. Much to his dismay, he found that legal expenses had doubled since the new senior vice president joined the staff. Careful analysis of legal costs and interviews with the other senior management staff determined that the new senior vice president saw more opportunities to use legal services than she could handle herself. She retained several new outside law firms to help with this ballooning legal workload, increasing green-dollar legal expenses considerably. This CEO eliminated the senior vice president of legal affairs position and learned a valuable lesson; he learned that carefully monitoring outside legal firm costs is absolutely crucial in strategic cost reduction.

MIDDLE MANAGEMENT CHALLENGES

Position titles for middle management vary widely in hospitals. A typical middle management title is "director." "Manager," "executive director," and "assistant vice president" are also common middle management titles. Small to medium-sized hospitals may have 25 to 50 middle managers. Large medical centers may have 100 or more.

The same questions a hospital must examine for its senior management must also be asked and answered for its middle management. Typically, there are ten times more middle managers than there are senior managers in hospitals. Is it really necessary? Often the candid answer is no.

Additional questions can be asked when specifically examining middle management. Are middle management positions occupied by highly competent people who have the professional skills necessary to provide leadership to their respective areas of responsibility?

Are the people occupying middle management positions competent to fulfill assigned duties, or are middle managers retained as a function of longevity and survival skills instead of professional competence? Is the hospital receiving real value from all its middle managers, or could some of the leadership functions be outsourced or delegated to consulting resources with better value?

MANAGEMENT STRUCTURE

How many layers of management are optimal in a hospital? Although no definitive guideline is available, hospitals that operate with fewer layers are more likely to have lower management costs. In small to medium-sized hospitals (annual net revenues less than $100 million) a maximum of four layers between the CEO and caregivers should be sufficient. In larger hospitals, a maximum of six layers should be sufficient.

Excessive layering of management in hospitals adds unnecessary costs. The costs of additional managers themselves are just part of this picture. Additional managers need support staff, such as assistants and secretaries. The additional managers take up valuable space and consume hospital resources. Hospitals with more than six layers of management, regardless of the hospital's size and clinical scope, are candidates for reducing management costs by reducing management layers.

Out-of-control layers

A medium-sized urban medical center began the process of evaluating its organization by drawing a physical organizational chart in the hospital's boardroom. The boardroom was used because it had the largest table in the building. The CEO and senior management team taped together a 15-foot by 8-foot blank sheet and proceeded to draw, by hand, the actual organizational chart that shows all layers of the hospital, from CEO to each employee in every department. The chart took every inch of the enormous blank sheet.

They were stunned by the completed chart because it showed up to nine layers of management in several areas of the hospital. For example, the layers between the CEO and delivery room nurse in the maternity unit were as follows:

1. CEO
2. Executive vice president/COO
3. Senior vice president of nursing
4. Vice president, specialty nursing
5. Assistant vice president, maternity
6. Director of labor and delivery
7. Assistant director of labor and delivery
8. Shift supervisor, nursing operations
9. Team leader, evening shift
10. Labor and delivery nurse

This exercise immediately prompted senior management to comprehensively evaluate all management layers throughout the hospital. Subsequently, management layers were reduced to six, eliminating a total of 30 management positions. Annual green-dollar cost savings of $2.5 million in direct salary costs and an additional $1 million in related costs for fringe benefits and support staff resulted.

HIGH COSTS OF INEFFECTIVE LEADERSHIP

Ineffective leadership increases hospital management costs in many ways. If ineffective leaders are in place at the senior level, the hospital's performance will generally be less than optimal. High expenses for middle management turnover associated with ineffective senior management also unnecessarily draw on the hospital's resources. Excellent middle managers rarely desire to work for an ineffective senior management superior. Costs associated with turnover of middle management disrupt the hospital's performance and add management green-dollar costs in the form of recruitment expenses. No hospital with an ineffective senior management team, an ineffective middle management team, or a bloated organizational structure can ever hope to be successful in strategic cost reduction.

ACTION PLAN CHECKLIST FOR REDUCING MANAGEMENT COSTS

Five simple but profoundly important factors influence a hospital's management costs. This five-step checklist for reducing management costs is an excellent guide for getting started with strategically reducing costs:

- *Step 1: Review organizational structure.* The CEO, along with trusted senior management, should review the hospital's organizational structure for excessive layering. An effective CEO should quickly identify layers of management or management positions that are no longer effective and are thereby adding costs to the hospital. The CEO should also be able to identify positions within middle management ranks that are ineffective or unnecessary, thereby adding costs to the hospital. These positions must be removed from the hospital's organizational structure permanently.
- *Step 2: Assess senior management.* Evaluate the personal performance of members of the senior executive team. Only the most competent and highest-performing senior executives should be retained if the hospital hopes to succeed in strategic cost reduction. Weak performers and persons whose positions do not fully contribute to the hospital's success should be let go as part of the strategic cost reduction process.

- *Step 3: Review middle management layers.* Once the senior management structure has been simplified, and members of the senior management staff have been personally evaluated for retention or termination, review middle management next. The middle management structure should have only those layers and positions that are required to provide leadership and guidance to the hospital's departments. Positions not justified should be eliminated to reduce hospital management costs.
- *Step 4: Assess middle management competence.* Once the middle management structure has been simplified, evaluate the performance of the individuals who occupy middle management positions. Again, only the highest-performing people should be retained. Weak performers or those whose effectiveness is no longer consistent with expectations should also be let go as part of the strategic cost reduction process.
- *Step 5: Support ongoing leadership assessment and development.* The most efficient hospitals are those that constantly assess senior and middle management structures as well as the people who occupy positions of responsibility. The highest-performing hospitals devote resources to leadership development on an ongoing basis. This is especially important for hospitals contemplating strategic cost reduction.

Some hospital executives might perceive these five steps to be overly basic. Some might consider the questions posed in these steps as having self-evident answers. If these steps seem too obvious or the questions are too easy, look in the mirror. The image you see is likely the principal reason management costs are too high in your hospital.

Strategic cost reduction begins at the top. Only when the hospital's management costs are as low as possible can the organization hope to achieve strategic cost reductions elsewhere. Assessing and reducing labor costs is the next area of opportunity in strategic cost reduction.

CHAPTER 6

Labor Costs

THERE ARE EXCELLENT opportunities for strategically reducing labor costs in hospitals. Employee productivity can range from extraordinary to dismal. Differences in labor productivity are indicators that labor costs in the least-productive hospitals can be significantly reduced without reducing quality. Quality of employees, versus quantity of employees, is an important labor-cost consideration. In some hospitals, the quality of employees is suboptimal, and opportunities abound for reducing costs by increasing the quality of employees. Further opportunities exist to better integrate hospital employees with support systems and technology to improve productivity while ultimately reducing labor costs. Permanent increases in productivity are an important element in the strategic cost reduction process.

Strategic cost reduction success is achieved with labor costs when employee productivity is high, quality of employees is high, and the synergy between labor costs and technology is such that the lowest costs of labor are achieved. Hospitals seeking to ensure that labor costs are as low as possible must strive to eliminate reasons for reduced productivity and eliminate excessive use of employee fringe benefits that do not contribute to the recruitment and retention of a quality workforce.

PRODUCTIVITY IMPROVEMENT OPPORTUNITIES

There are many opportunities to improve productivity throughout the hospital. Productivity is at its absolute highest when the work being performed is the only work required to fulfill that certain task. Additional work performed outside the required task increases labor costs and reduces productivity. Work that merely fills the day versus work that achieves a patient care or customer service outcome certainly increases costs, but not necessarily quality. Sharing workload tasks between departments improves productivity while reducing labor costs. On the other hand, when no workload sharing or workload integration takes place, costs can be unnecessarily high. The presence of technology support for hospital staff improves productivity and reduces costs. Hospitals intending to achieve strategic cost reduction success must pay careful attention to productivity improvement throughout the organization.

Smart sharing

A small rural hospital struggled to balance patient care needs with its obligation to be soundly managed financially. During a strategic cost reduction initiative, every job on every shift was examined by directors and employees for opportunities to consolidate duties and reduce labor costs. Night-shift employees in the admitting department and the emergency department identified one such opportunity.

The admitting department was staffed with one employee—a registrar—on the night shift, even though the number of patient admissions was minimal. The emergency department night shift was staffed with two nurses and a unit secretary to answer the phone and handle supply and support activities. Neither the admitting department employee nor the unit secretary for the emergency department was fully engaged and productive during

> the shift. By combining positions, green-dollar labor costs were reduced by $35,000, and the remaining position was much more interesting and challenging. Similar consolidations took place elsewhere in the hospital, enabling it to reduce its workforce by 5 percent in total, placing it solidly in the black for the first time in a decade.

Hospital labor productivity is usually measured by the ratio of paid full-time equivalent employees (FTEs) per adjusted occupied bed (AOB); the higher the ratio, the lower the productivity. Some hospitals can provide excellent quality service with high productivity levels, such as 3.5 FTE/AOB. Other hospitals struggle with labor productivity, running at a low level of productivity, such as 7.0 FTE/AOB. Regardless of the productivity starting point, all hospitals have opportunities to improve labor productivity performance. Expecting employees to work hard, collaborate with fellow employees, avoid excessive time off, and use technology whenever possible are all common sense and highly effective methods to improve employee productivity and lower labor costs.

STAFFING MIX AND TECHNOLOGY

The mix of professional staff, technical staff, and support staff is an important consideration in maximizing productivity and minimizing hospital labor costs. If the professional mix is inappropriately weighted toward more expensive staff, labor costs will be higher than necessary. Certain hospital departments have a strong tendency to overweight professional staff and underweight technical and support staff. For example, it is common in hospital-based skilled nursing facilities to overuse registered nurses and underuse patient care technicians or nursing assistants. The acute care orientation of hospital-based skilled nursing units differs from the orientation of well-managed nursing homes that care for the same types of patients. Well-managed nursing homes generally have far fewer registered nurses and far more technical staff than hospital-based skilled nursing units do. Well-managed nursing homes typically have lower labor costs to take care of the same number and same kinds of skilled nursing patients.

Therefore, each hospital department should carefully evaluate its staffing mix and periodically make changes to ensure that the lowest possible labor costs are achieved while the expected quality of service is delivered. The mix of professional, technical, and support staff should be constantly adjusted as the department's services change and as technologies change. For example, as electronic medical records systems are implemented, the need for trained medical record coders is significantly reduced. As PACSs (picture archiving and communications system) are implemented in radiology departments, the need for radiology technicians is significantly reduced. Such technology advances affect nearly every hospital department. Constant monitoring of technology and adjustment of staffing mixes are necessary to ensure that labor costs are as low as possible.

LOST PRODUCTIVITY

There are many opportunities for employee productivity to diminish in hospitals. Lack of job skills can sap hospital labor productivity. Overuse of paid time off, such as sick time or disability, can reduce productivity. Clock-watching behavior, if employees check in late and plan to leave early, can also reduce productivity; such behaviors diminish an employee's actual on-the-job time.

Lost productivity because of lack of competence significantly increases hospital labor costs; if an employee is not skilled in her job requirements, many different costs can be unnecessarily increased. For example, an unskilled radiology technologist who constantly has to repeat examinations causes an increase in film costs and a delay in the patient's treatment. Employees who are continuously on sick leave or disability usually add to labor costs because overtime or temporary staff have to be used. Lost productivity through lack of job skills or overuse of paid time off present leadership challenges to middle managers. When middle managers who directly supervise employees address productivity problems, labor costs can be lowered as much as possible. When these problems are ignored, strategic cost reduction success cannot be achieved.

FRINGE-BENEFIT COST CHALLENGES

Fringe benefits account for as much as 30 percent of the total labor costs for hospitals. Health insurance is the largest component of fringe-benefit costs for hospital employees. Labor costs are unnecessarily inflated when hospital employees pay higher premiums for health insurance than their counterparts do in other industries. The absence of incentives for wellness can contribute to the high cost of healthcare utilization by hospital workers. Ironically, few hospital leaders will insist that employees maintain healthy weights, discontinue smoking, or diminish otherwise unhealthy habits. Every dollar spent taking care of a hospitalized healthcare worker whose unhealthy habits have ultimately caught up with him is a green dollar wasted.

Hospitals can strategically lower labor costs by carefully considering how fringe-benefit green dollars are spent. One middle-sized community hospital observed that its workers' compensation costs were excessive because of an inordinately high rate of employee back injuries. Careful study and extensive interviews with injured employees led to the discovery that the hospital had not kept up its training programs for lifting patients nor had they invested in assisted lifting devices to aid employees. Addressing both of these shortcomings led to a 50 percent decrease in workers' compensation claims the following year, immediately saving the hospital more than $100,000.

EXCESSIVE EMPLOYEE TURNOVER COSTS

When a hospital experiences excessive employee turnover, productivity is deflated and labor costs are inflated. Recruiting costs to conduct replacement searches for workers who have left the hospital are exorbitant. Costs associated with temporarily filling open positions are often extraordinarily high because they necessitate using in-house staff to work overtime or temporary workers who are supplied by outside agencies. Cross training and orienting new employees also add to the overall cost of labor in a hospital setting.

> **Turnover trouble**
>
> Over a three-year period, a small community hospital's pharmacy department experienced a 50 percent turnover rate for pharmacists. Recruiting costs were extremely high, and the hospital had to rely on agency pharmacists to fill vacancies at a pay rate triple the rate of employed pharmacists. During a strategic cost reduction initiative, pharmacists were encouraged to provide candid input regarding turnover problems and possible solutions.
>
> Rising to the challenge, pharmacists communicated to senior management that local retail pharmacies had higher pay rates, more flexible work schedules, and better access to computer technology for identifying potential medication conflicts. Working with its remaining pharmacists, the hospital was able to restructure the compensation and work schedules. The hospital also invested in an upgraded pharmaceutical information system that helped pharmacists identify potential problematic medication conflicts. The turnover rate for pharmacists dropped to 20 percent the following year, saving the hospital more than $150,000 in recruiting costs and temporary agency pharmacist fees.

Employee turnover should be decreased whenever possible. Any hospital department that experiences turnover greater than

10 percent per year should carefully evaluate and address the reasons for the excessive turnover. Senior management should be particularly concerned if the annual turnover rates exceed 10 percent in any department; they should intervene as necessary to ensure overall hospital employee turnover rates are lowered whenever possible. If excessive employee turnover is attributable to an incompetent middle management team, senior management should be especially concerned. If good hospital employees are seeking employment elsewhere because of conflicts with an incompetent middle manager, there is no excuse for retaining that middle manager's services.

INAPPROPRIATE RETENTION COSTS

Just as high turnover can increase labor costs, so can excessively low turnover if this means incompetent employees are being retained. Inappropriate retention of employees with poor clinical skills, poor customer service skills, or poor attitudes can lead to excessively high labor costs. How can this be true? Hospitals that neglect to let poor clinical performers go experience increased costs in several ways. First, coworkers must defer their primary duties to address the poor performer's shortcomings. Second, a poorly skilled staff member's mistakes inevitably lead to repeating services, which increases costs and inconveniences patients.

Another fundamental labor-cost challenge is personal productivity. Some employees just do not work very hard, and hospitals accept this poor work ethic. Even though it may not be politically correct to say, it is nevertheless true that hospitals employ some lazy people. When this is the case, coworkers must accommodate for the lack of efficiency. Terminating the employment of unproductive workers with poor work ethics will most certainly reduce hospital labor costs and improve employee productivity. Employee morale may improve too.

Rotten apples

A community hospital engaged in strategic cost reduction received an unanticipated proposal from the respiratory care department. This department, which comprised ten employees, was plagued with excessive turnover, employee unrest, and incessant customer service complaints from physicians and patients. When the CEO asked all employees to be creative in recommending green-dollar cost reductions, respiratory care employees took the request seriously.

Eight of the ten employees offered to work harder and change work schedules. They even offered to consider a pay cut if two of their coworkers could be transferred elsewhere or be terminated. It seemed that two employees had long histories of disruptive behavior and poor work ethic. They were the classic "rotten apples" who spoiled the barrel.

The CEO agreed to terminate the two identified employees; their personal files had ample evidence of long-term problematic behavior. The remaining eight employees shared the department's duties and eliminated patient and physician complaints of poor service. Green-dollar costs were reduced by $90,000 per year after the two employees left the department, and the department director was counseled about her accountability for the department's problems. She eventually became a much stronger department leader and one of the hospital's biggest middle management advocates for strategic cost reduction.

OUTSOURCING CONSIDERATIONS

In recent years, hospitals have pursued outsourcing with great vigor, believing outsourcing will lower total labor costs in some operational areas. However, outsourcing should only be considered when it can be definitively demonstrated that costs will, indeed, be lower. In specific areas that require repetitive work, such as medical records transcription or billing follow-up calls, outsourcing may be an ideal solution. Outsourcing companies can standardize, recruit, and retain staff more easily than hospitals can. They may even take certain jobs to other countries where labor costs are considerably less. Hospital leaders should be aware, however, that hidden costs are sometimes associated with outsourcing. This is especially true if costs and quality are reduced simultaneously.

ACTION PLAN CHECKLIST FOR REDUCING LABOR COSTS

Labor costs represent an immense opportunity for hospitals desiring to strategically lower costs. Following are several key strategies to begin this process:

- *Step 1: Set productivity target.* Hospitals should set an overall productivity target that is consistent with their size, complexity of services, and reimbursement environment. They should then function within that target. Small to medium-sized hospitals (annual net revenue less than $100 million) should target 4.0 FTE/AOB or better. Larger hospitals should target 5.0 FTE/AOB or better. Large medical centers should never exceed 6.0 FTE/AOB.
- *Step 2: Review professional mix.* Hospitals should establish professional-mix parameters for each department that are consistent with achieving quality patient outcomes and incurring the least overall cost. This mix should be reevaluated constantly in light of changes in technology.
- *Step 3: Set high Standards for the workforce.* Hospitals should set the highest standards for performance and work ethic for their workforce because they can correspondingly reduce labor costs. Department directors are in the best position to evaluate work ethic and job skill performance and should be expected

to set a positive personal example by working diligently and effectively themselves.

- *Step 4: Fire poor performers.* Hospitals should not retain employees who have productivity problems, skill-quality problems, or customer service problems. Suitable training of management at all levels in how to evaluate employee performance and how to let poor performers go is essential to achieve success in this strategy.
- *Step 5: Minimize turnover.* Hospitals should establish leadership standards that clearly focus on minimizing employee turnover and maximizing retention of excellent employees. A turnover rate greater than 10 percent in any department is cause for concern and intervention.
- *Step 6: Actively manage fringe benefits.* Hospitals need to pay strict attention to the design of fringe-benefit programs to achieve the lowest benefit costs that are consistent with recruiting and retaining excellent employees.
- *Step 7: Make careful outsourcing decisions.* Hospitals should outsource only those activities that clearly reduce hospital green-dollar costs and preserve or enhance quality and customer service.

Hospitals with effective leadership will achieve high levels of employee productivity while simultaneously experiencing the lowest possible labor costs. Successful strategic cost reduction depends on achieving and maintaining high levels of employee productivity. Once labor costs are as low as possible and consistent with quality standards, hospital senior and middle management teams can focus on reducing supply and service costs.

CHAPTER 7

Supply Costs

HOSPITALS HAVE EXCELLENT opportunities to strategically reduce the cost of supplies, which range from high-tech clinical supplies to low-tech office supplies. Comfort with the status quo is too often the rule of thumb when purchasing hospital supplies. Some hospitals fail to achieve the lowest possible supply costs because they do not use effective negotiating skills when purchasing required supplies. Aggressive evaluation of alternative supplies, prudent and effective purchasing of supplies, and collaboration between hospitals and physicians are effective approaches to strategically lowering hospital supply costs.

Although hospitals use a massive amount of supplies, keys to success for strategically lowering supply costs are simple. Opportunities abound to strategically reduce hospital supply costs with no negative effect on clinical quality or service to patients. The essence of the challenge for hospitals is to purchase only supplies needed at the lowest possible green-dollar cost. Eliminating waste and considering supply alternatives are further keys to success.

SUPPLY SELECTION CHALLENGES

Selecting the appropriate supply for every application is the first step in ensuring that hospital supply costs are as low as possible. Using a suture kit that has 12 components for a surgery that only needs 6 of those components obviously wastes supply dollars. There are tens of thousands of supply applications in a hospital setting. Making sure the right supply is at the right place, at the right time, with the right purpose can be a challenge. Ignoring any one of these factors can needlessly increase supply costs.

> **Miracle drug**
>
> As part of a strategic cost reduction initiative, nurses, pharmacists, and physicians within the hospice unit of a major community hospital researched medication options for terminally ill patients. The research led the unit to an alternative antinausea medication for chemotherapy patients. They discovered an alternative to the currently used drug that was more effective in reducing nausea and further reduced green-dollar costs by $18,000 per year. This "miracle drug" discovery led other clinical departments at the hospital in conjunction with physicians to search for medication alternatives. In total, the hospital implemented alternatives that saved $250,000 per year with no untoward clinical implications. Not only was quality of care preserved, but, in some cases, quality was actually improved by the use of the alternative, less costly medications.

Ensuring that a specific supply will have its desired effect is also a challenge. For example, an antibiotic that has a broad spectrum of implications may be used when a less expensive, narrow-spectrum antibiotic is sufficient. This wastes supply dollars and may contribute to a growing public health problem concerning bacteria resistance to antibiotics. Being sure that the supplies for the operating room are selected for their intended purpose versus being selected because a physician personally prefers them is another challenge.

The quality versus benefit equation is another important consideration when evaluating supply costs. Providing Mercedes quality when Chevrolet quality is sufficient needlessly wastes supply dollars. The use of different contrast media for certain imaging procedures is an excellent example. One material offers maximum contrast for a wide range of procedures. Another material, which is significantly less expensive, may not have the full-scale versatility the first one has but is just as effective for certain radiology procedures. For these procedures, use of the alternative contrast media is less costly and has no adverse clinical implications.

The dilemma of using disposable supplies versus reusable supplies is another factor in the supply-cost equation. In recent years, hospitals have started returning to reusable supplies after discovering that the exclusive use of disposables not only is more expensive but can be harmful to the environment as well. The use of reusable pulse oximeter probes in place of disposable probes, for example, has been demonstrated by some hospitals to save green dollars without compromising quality. Green supply purchases can save green dollars.

COLLABORATING WITH PHYSICIANS

Physicians have a profound impact on the selection and utilization of hospital clinical supplies. Involving physicians in a true and objective evaluation of supply options is challenging, however. Often, physicians desire to continue to use a specific supply, drug, or medical appliance simply because they have used it in the past with good results. It is the hospital's obligation to collaborate with physicians to enable them to truly evaluate supply options and to jointly select those supplies, drugs, and other material that fulfill their intended purpose while minimizing green-dollar supply costs.

This approach is especially challenging when recognizing the aggressive and even unscrupulous practices of some medical supply companies. In the past, it has not been uncommon to discover that suppliers have provided incentives, bordering on bribes, to encourage physicians to select or continue using their particular supply. Hospitals must be conscious of these practices and ensure that countermeasures are in place to guard against this inappropriate selection of supplies.

Yet another challenge in the effective purchase of supplies is avoiding supply sacred cows. If a department manager, physician, or even an employee has a special preference for a particular supply and is not willing to evaluate alternatives, the continuing purchase of these sacred supplies will undoubtedly and inevitably raise supply costs.

> **Conflicted cardiologist**
>
> The CEO of a community hospital who was engaged in a strategic cost reduction initiative learned first hand the power of a sacred cow run amok in the boardroom. This hospital's leading cardiologist was a member of the hospital's board of directors. Even though the board had approved the strategic cost reduction initiative, the cardiologist resisted all attempts of the hospital to jointly consider alternative options for purchasing pacemaker devices. The cardiologist steadfastly refused to evaluate any brand other than his costly preference, citing quality considerations.
>
> The hospital independently pursued the evaluation of alternatives with help and support from the cardiology department of a nearby academic medical center. With this assistance, the hospital was able to select a new supplier for pacemakers, which saved more than $150,000 per year in green-dollar supply costs. Grudgingly, the hospital's leading cardiologist agreed to use the pacemakers from the new supplier. He later confided in a physician colleague that he would, sadly, not be getting any more all-expenses-paid ski trips from the hospital's original supplier of pacemakers.

The overuse or waste of supplies is also a challenge with physicians. In the operating room especially, it is not uncommon to waste supply packs by using only four or five items in a pack that contains 10 or 15 items that are not really necessary for the particular surgical case. Ultimately, these practices waste supply dollars and increase the overall cost of hospital care.

EFFECTIVE PURCHASING

Selecting the right supply is important, but purchasing the right supply effectively is even more important. Strong, competitive bidding and thorough evaluation of alternative supplies are the cornerstones for ensuring that supply costs are as low as possible. Indeed, evaluating alternatives for clinical and nonclinical supplies is essential to run a truly competitive bid process for supplies.

Hospitals that participate in group purchasing arrangements sometimes assume the group purchasing organizations are getting the lowest cost and the best selection of supply products. Although that is often the case, it is still critical for individual hospitals to periodically evaluate both the quality of supplies and the price of supplies to ensure that they are getting the best possible value. Joining a group purchasing organization is not an excuse for paying a higher price for supplies.

> **Negotiation home run**
>
> In collaboration with physicians, a large medical center embarked on a strategic cost reduction initiative in its surgery department. A committee of operating room managers, leading surgeons, and nurses identified the most expensive orthopedic supplies, prostheses, interoperative drugs, and procedure packs and then systematically evaluated their alternatives. Physicians provided input on quality variances and identified multiple supply options that

> met quality standards. Managers then used competitive bidding to ensure fair and aggressive competition between vendors. As a result, deep discounts in all supply categories generated green-dollar cost reductions of more than $500,000 per year for orthopedic supplies.

Inventory management is another key factor in strategically reducing supply costs. The use of "just in time" inventory procedures has helped hospitals reduce their overall inventory carrying costs dramatically. Continued efforts to rotate obsolete inventory items out of hospital inventories will surely reduce supply costs in the future.

ACTION PLAN CHECKLIST FOR REDUCING SUPPLY COSTS

Strategically reducing supply costs is a major commitment that requires multiple pathways to an effective implementation. The following steps are essential:

- *Step 1: Develop a comprehensive master supply list.* Delineate every supply item the hospital uses within each individual department. The type of supply, the annual quantity used, and the person or department responsible for purchasing the supply should be included. This master supply list should be updated continuously.
- *Step 2: Consider alternatives for review.* Once all supplies the hospital uses are definitively listed, it is the responsibility of each department to identify at least two or three alternatives for every supply.
- *Step 3: Evaluate supply performance.* Users of the supplies should next evaluate the quality, efficacy, and cost of supply

alternatives. Evaluation comments should be valued and taken into consideration when making the final supply-selection decision. Physicians should be intimately involved in evaluating clinical supplies.

- *Step 4: Ensure effective purchasing.* When users have selected the desired supply items, the next and most important step is to purchase those items at the lowest possible cost. Competitive bidding is a requirement to ensure that the hospital is not paying a higher price than necessary. Involving upper management in supply purchases is also an important strategy. For high-cost supply items (greater than $50,000 per year) a vice president–level member of the hospital's senior management team should be personally involved in the supply purchasing process.
- *Step 5: Select the lowest-cost/highest-value supply.* The ultimate purpose of critically evaluating supply alternatives and effectively purchasing the supplies is to select the items that perform the needed function at the lowest possible cost. Only after each and every supply is comprehensively evaluated is it possible to determine whether or not the lowest possible cost is being achieved for every supply.

Supply costs are a critical factor in strategic cost reduction. Purchase of services by the hospital is another important aspect of strategic cost reduction success. Service purchase strategies are discussed in Chapter 8.

CHAPTER 8

Service Costs

HOSPITALS PURCHASE ENORMOUS amounts of services from outside organizations. Each of these services represents an opportunity for strategic cost reduction. In some instances, there are opportunities to reduce costs by more prudent purchasing practices and more effective negotiation strategies. In other instances, service costs can be reduced by acquiring technology to reduce dependence on purchased services. In still other instances, service costs can be reduced by outsourcing or insourcing practices. Effective management of purchased services and periodic review of alternatives can identify excellent opportunities to achieve strategic cost reduction success.

What services do hospitals purchase? They purchase maintenance contracts for high-tech imaging equipment and low-tech elevators. They purchase laundry services from across town and medical transcription services from India. They purchase legal services and financial audit services. They contract for emergency physician coverage and pathology services. In essence, hospitals purchase thousands of services that they cannot provide themselves because of capital, economic, technological, or facility constraints. Strategic cost reduction success requires that services are purchased at the lowest green-dollar cost and the highest quality appropriate to perform the hospital function.

SERVICE OPTION CHALLENGES

Hospitals use purchased services to achieve certain operational or clinical outcomes. Purchasing the right services, in the right quantity and at the right time, is key to ensuring that the costs for purchased services are as low as possible. It is vital for hospitals to periodically consider options for all purchased services. For example, when purchasing accounting and legal services, hospitals may fall prey to past-practice inertia and rarely, if ever, consider options. Sometimes options are not considered because of conflicts of interest or sacred cow situations with physicians or even hospital board members. In other instances, hospitals may be lulled into a false sense of security with group purchasing arrangements, assuming incorrectly that the lowest possible service costs are always being delivered. All hospitals should rigorously consider options when purchasing services of any kind.

Sacred decorator

While a large community hospital was engaged in strategic cost reduction, department directors staged a mini-revolt. The CEO made it clear that strategic cost reduction initiatives would not be thwarted by sacred cows. Department directors took his words to heart, and years of frustration with the hospital's contracted interior decorator boiled to the surface. Department directors carefully made a list of examples of "decorator excess" and presented it to the CEO.

The list was highly embarrassing. The interior decorator, over the rigorous objections of the director of housekeeping services, had insisted on purchasing a set of $800 trash cans for the hospital lobby. Over the objections of the nursing director, 20 new chairs had been selected for the intensive care unit (ICU) waiting room that were completely unsuitable for elderly visitors. A floor-covering material was chosen for the rehabilitation department that proved to be dangerously slippery for disabled patients, although beautiful in appearance.

> The interior decorator had been retained for a decade at the recommendation of the chair of the hospital board of directors. For ten years she had exercised dictatorial power in selecting furniture, color schemes, and even trash cans. Until the strategic cost reduction initiative emboldened directors, no one felt empowered to challenge her authority. Subsequently, a different interior design firm, with more experience in the healthcare field, was selected to replace the "sacred decorator." The green-dollar cost for the new contract service was much lower, and $800 trash cans were never purchased again.

COLLABORATION WITH PHYSICIANS

Physicians have a vital interest in purchasing the services directly related to clinical practices. For example, they may prefer certain reference labs for outside clinical testing. In other instances, physicians may prefer an outsourced emergency medicine service because of years of positive experiences. Overcoming inertia, momentum, and adherence to past practices is a purchasing services challenge related to the medical staff. In addition to inertia, sacred-cow issues with certain purchased services may preclude management from truly considering alternatives. However challenging, it is critical that physicians are meaningfully involved in evaluating alternatives for any purchased service with clinical implications.

CAPITAL INVESTMENT OPPORTUNITIES

In some cases, hospitals will incur excessive purchased service costs because of their inability to make capital investments. For example, acquiring a specialized piece of laboratory testing equipment may be costly initially, but eliminating a long-standing reference laboratory service is likely to more than offset the equipment's initial capital costs. Hospitals should be constantly evaluating opportunities to substitute capital investments for outsourced purchased services. It is the responsibility of senior management and middle management to be consistently vigilant for these opportunities and to rigorously pursue these alternatives.

> **Rent versus buy**
>
> A large community hospital involved in evaluating purchased service costs carefully compared the rental costs of specific clinical equipment, such as pumps and special trauma patient beds, with the capital investment costs of purchasing those pumps and beds. After analyzing rental costs for the previous three years, the hospital determined that it should invest $500,000 to purchase enough pumps and special purpose beds to avoid the necessity of renting them 90 percent of the time. The hospital was able to reduce rental fees by $350,000 per year by paying for the new pumps and beds in 17 months and saving $29,000 per month in green-dollar service costs thereafter.

OUTSOURCING

In some instances, outsourcing may present opportunities to strategically lower service costs. For example, if an outsourcing vendor can provide a service for a lower cost because of efficiencies of scale and consolidation of technology, using an outsourcing approach may strategically lower costs. In other instances, it may be possible to outsource services to improve quality while keeping costs at the same level or lowering them. In yet another situation, outsourcing can provide access to consolidated labor pools less expensively than a hospital can acquire them independently. Outsourcing opportunities exist for virtually every service that a hospital purchases. It is up to senior and middle management to constantly evaluate opportunities and determine which opportunities can strategically lower service costs while maintaining or improving clinical quality and customer service.

INSOURCING

As the opposite of outsourcing, insourcing may also present opportunities to strategically lower service costs. Sometimes services are outsourced because of the hospital's inability to manage a particular challenge. For example, housekeeping services are outsourced because of the hospital's inability to retain a good executive housekeeper or a stable workforce of housekeepers. Over time, situations change. Perhaps the initial reason the hospital was unable to retain a good executive housekeeper or stable workforce has changed. In that circumstance, it may be possible for the hospital to bring the housekeeping services back in house and lower service costs at the same time. In this sense, the hospital is able to lower service costs by practicing an insourcing strategy.

NEGOTIATING OPPORTUNITIES

When a hospital purchases services, there are almost always opportunities to negotiate better prices with the service vendor. This is especially true in instances where effective negotiation was not historically present between the hospital and the service vendor. The first step in negotiating lower costs is to identify and evaluate realistic service-delivery alternatives. It is difficult to negotiate a better price if a vendor believes the hospital will never change the service being provided. On the other hand, if the vendor is constantly and constructively reminded that the hospital is considering other service choices, the vendor is more likely to offer the best possible price.

In addition to evaluating service alternatives, using the "position power" of senior executives is another consideration in obtaining the lowest possible purchased service cost. It is not uncommon for hospitals to use middle managers for negotiating even large purchased-service contracts. If service vendors know they are dealing exclusively with middle management, it is unlikely they will deliver the best possible deal. On the other hand, if senior management executives, those at the vice president level for instance, are involved in the negotiating process, the hospital has a much better probability of obtaining the best possible deal.

> **COO success**
>
> A regional tertiary hospital was involved in a comprehensive review of all service purchases. One of its largest-dollar service contracts under review was in information systems maintenance. During a strategic cost reduction initiative, the COO became personally involved in the contract negotiation process for all service contracts in excess of $100,000. He demanded personal negotiations for the new contract with the service company's CEO. After initially refusing to negotiate, the service company agreed to a 10 percent reduction in the new contract, saving $200,000 per year in green dollars over the previous contract. Before the COO's interjections, no one above the department director level was ever involved in negotiating this $2 million per year service contract.

COMMON-SENSE OPPORTUNITIES

As with lowering supply costs, lowering service costs can be achieved through common sense. The person closest to the service purchase decision is the best person to apply this common sense. Consider the example of a hospital facilities manager who evaluated service contract costs for the hospital's elevators. He determined that he did not need a 24-hour maintenance contract because the service vendor was located only minutes away and was always available on call. By switching to an on-demand service arrangement, the cost of elevator maintenance was reduced by 75 percent with no reduction in quality. In many ways, opportunities abound for common-sense applications in the purchasing of services in a hospital setting.

> **Courier abuse**
>
> An urban community hospital evaluated many opportunities to make common-sense reductions in service costs. One of the best ideas came from an employee in the mailroom. She encouraged a comprehensive evaluation of the use of courier services. When the service was introduced a decade earlier, it was used only for urgent needs, such as delivering blood samples for a critically ill patient to a regional referral laboratory or picking up a highly specialized medical device for an emergency operation.
>
> What transpired over the years, however, was that the courier service came to be used more and more for nonurgent pickups and deliveries. At the time the service was reevaluated, it was determined that 90 percent of courier use was for nonurgent items, like delivering hospital mail and memos to physician offices. As a result of the reevaluation, strict controls were put on the use of the courier service, which reduced green-dollar costs by 90 percent.

MANAGING PURCHASED SERVICES

A most important consideration in actively managing purchased services is the dedicated, periodic consideration of alternatives for each and every service purchased by the hospital. Past-practice momentum is the enemy of strategic cost reduction in service purchases. When options are actively considered and evaluated, it is much more likely that green-dollar costs will achieve their lowest possible level and be sustained there. A middle manager should be accountable for evaluating every purchased service and considering alternatives on at least an annual basis. In addition, senior management executives should be accountable for actively evaluating the service purchases they directly control to ensure that the lowest-cost options are in place.

> **Audit inertia**
>
> A major tertiary hospital had used the same "big three" audit firm since the firm had been one of the "big six," two decades previously. Many hospital board members had professional ties to this favored audit firm; the hospital's CFO was actually a former partner. As part of the strategic review of all service contracts, the hospital's financial audit was put for competitive bid using a formal request-for-proposal process. At the end of the process, the hospital selected a new audit firm, a regional firm instead of one of the national big three firms. Savings in green dollars exceeded $50,000 per year.
>
> During the first year with the new audit firm, it became clear to both the board and senior management that the regional firm did a much more thorough job on the audit for a much lower price than the big three audit firm had done. One of the noticeable differences was in the management letter's detailed questions that were presented to management at the conclusion of the audit. No one could escape the realization that the highly comfortable relationship with the previous audit firm resulted in less aggressive questioning of management. In this case, the hospital won on all counts. It achieved lower service costs and better service quality, and it eliminated a conflict-of-interest problem with the CFO and several board members.

ACTION PLAN CHECKLIST FOR REDUCING PURCHASED-SERVICE COSTS

Reducing purchased-service costs is a monumental commitment that requires constant vigilance and a multistep process to implement effectively. The following steps are essential:

- *Step 1: Develop a comprehensive master service list.* The type, quantity, and annual costs of each purchased service should be delineated and constantly updated. A responsible manager should be assigned to monitor and oversee each purchased service.
- *Step 2: Consider alternatives for evaluation.* For each purchased service, there should be a continuously updated list of alternatives for the service. The person responsible for monitoring the purchased services should periodically evaluate each alternative and determine if changing service vendors or approaches will lower costs and/or improve quality. In addition to evaluating service vendor options, consideration should be given to evaluating capital investment or technology substitutions to determine if the service purchased could be eliminated, either partially or completely.
- *Step 3: Evaluate effect on end users.* When considering alternatives, the hospital should constantly evaluate the effect on the actual end users of a purchased service. The overall goal should be to strategically lower service costs as much as possible while simultaneously improving clinical quality.
- *Step 4: Ensure effective purchasing.* Senior and middle management are responsible for evaluating service alternatives, obtaining competitive bids, and negotiating the lowest green-dollar service costs, on at least an annual basis, for each service purchased by the hospital.
- *Step 5: Select the lowest cost/ and highest value.* It is the responsibility of senior management to ensure that all purchased services represent the lowest green-dollar cost and highest value to the ultimate end user.

When management, labor, supply, and service green-dollar costs are as low as possible, hospitals can find further strategic cost reduction opportunities in the area of clinical utilization, which is the subject of Chapter 9.

CHAPTER 9

Clinical Utilization Costs

SOME OF THE BEST untapped, yet most challenging, opportunities for strategic cost reduction are in the area of clinical utilization. Collaboration between the hospital and physicians can collectively ensure that clinical services are delivered to patients in the most efficient manner and at the lowest cost possible. Hospital leaders are responsible for providing an environment in which physicians can diagnose and treat patients in a manner consistent with quality and cost needs. Efficient clinical utilization can contribute to reducing the variability of care, reducing error-related costs, and facilitating more efficient use of diagnostic services and shorter lengths of inpatient hospital stays. Managing the overall process of patient care in the most cost-efficient manner is the mutual responsibility of physicians and hospital leadership.

COLLABORATION BETWEEN HOSPITAL LEADERS AND PHYSICIANS

The hospital's role is to foster an optimal environment that simultaneously facilitates efficient use of resources and high-quality clinical care. When hospital leaders and physicians collaborate, an optimal environment of care can be created and sustained. The optimal environment for care makes it possible for physicians to admit, treat, and discharge patients in the most efficient manner. Eliminating unnecessary tests and clinical procedures, as well as providing the shortest possible length of stay for inpatients without sacrificing high-quality care, leads to strategic cost reduction success. Hospital leaders must be attentive to barriers that can reduce physician efficiency, such as slow turnaround time for laboratory and radiology tests, problematic scheduling of operating rooms, and unreliability of hospital services on weekends, to name just a few. Constant communication between hospital leaders and physicians on matters of clinical utilization is essential to identify barriers and eliminate them promptly.

New math

One community hospital discovered that adding labor costs actually decreased overall costs in the surgery department. A collaborative evaluation of ways to reduce between-surgery case turnaround time that involved operating room nurses and surgeons led to a nonintuitive conclusion. Case turnaround time was at its worst in the late afternoon, which was after day-shift nurses and surgery department housekeeping staff had left for the day.

When late afternoon surgery cases were added to the schedule, on-call nursing and housekeeping staff were called in. This added to delays and costs because on-call pay and overtime pay became part of the total green-dollar cost equation. A review of the number of late afternoon cases confirmed that it would be more cost effective to add a surgery and housekeeping crew through 9 p.m. After implementing this strategy, the time between cases dropped dramatically and eliminated overtime and on-call costs paid for the additional labor costs of the new late-afternoon operating room and housekeeping crew.

COSTS OF VARIABILITY OF CARE

In an ideal world, all patients with the same disease or clinical problem would be treated consistently. The concept of clinical pathways was created to facilitate this consistency. When care varies from an ideal norm, costs rise and quality may deteriorate. Overuse of diagnostic tests because of physician variability and personal preferences contributes to higher-than-necessary costs, just like the overuse of treatments and clinical procedures leads to higher-than-necessary costs. On the other hand, consistency in the use of diagnostic tests, clinical procedures, supplies, and patient services enables the hospital to anticipate patient needs and deliver care efficiently. The higher the level of consistency, the lower the costs. The higher the variability in patient diagnoses and treatments, the higher the costs.

HIGH COST OF ERRORS

Hospital errors, no matter their genesis, increase clinical utilization costs. Hospital errors require the repetition of diagnostic procedures and therapeutic treatments and unnecessary extension of inpatient hospital lengths of stay. The collective cost of hospital errors is profoundly high. Reducing these errors to the lowest possible level allows strategic cost reduction goals to be achieved. Clinical errors, such as performing surgery on the wrong site, diminish positive patient outcomes and are extraordinarily costly in all respects. Hospitals that focus on reducing errors of all types have a much better chance of achieving strategic cost reduction success.

Errors can increase costs directly and indirectly. Hospital-acquired infections are thought to cause more than 100,000 deaths per year and countless extended hospital stays. One in 20 hospital inpatients acquires an infection during a hospital stay. The costs of these infections alone are enormous, approximately $30 billion annually, not to mention the costs of lesser errors (McCaughey 2006). Malpractice suits and higher costs for liability insurance add further to hospital costs.

CLINICAL OVERUTILIZATION AND EXTENDED LENGTHS OF STAY

It is widely reported that defensive medicine increases healthcare costs. Why else would a pediatrician prescribe an antibiotic for what is suspected to be a viral infection? The pediatrician will often submit to parental demands and prescribe an antibiotic even if he knows it will do no good and may potentially even do harm. This increases clinical utilization costs and may even cause a potentially deadly resistance to antibiotic treatments for children in the future.

Every day a patient stays in an inpatient hospital bed versus being discharged into a less-intense clinical setting or her home unnecessarily increases the cost of that patient's care and the overall cost of the hospital. Therefore, overusing clinical services and extending inpatient hospital stays, whatever the root cause, increase the costs of healthcare dramatically.

MANAGING CARE OPPORTUNITIES

Hospital leaders and physicians have many opportunities to work together for the purpose of strategically lowering costs. This is commonly called "physician alignment." Alignment opportunities may also improve clinical outcomes. Utilization costs may be significantly reduced, according to the extent that hospital leaders and physicians can collaborate. Every hospital should have a solid interface for hospital leaders and physicians to work together to generate opportunities for decreasing utilization costs.

> ### Ad hoc committee success
>
> A busy community hospital was experiencing long delays in getting emergency department patients into intensive care because of high census and low patient turnover in their ICU. It was not

uncommon for critically ill patients to be held in the emergency department for 24 hours awaiting an ICU bed. This multifaceted problem created a perfect storm of low quality and high costs. Quality was lower than desirable because patients could not receive the critical care they needed while being held in the emergency department setting. Costs were higher than necessary because the delays in initiating critical care intervention, only possible in the ICU, caused unnecessary extended hospital stays.

The medical staff president responded by appointing an ad hoc committee of emergency department physicians and nurses; critical care physicians and nurses; and ancillary department representatives from radiology, laboratory, and respiratory care. He charged the committee with coming up with a permanent solution in 30 days. The committee quickly identified clinical and communication bottlenecks that were causing patients to be held longer than necessary in ICU. Physician communication problems and delays in diagnostic tests were the main culprits. New expectations and communication protocols were immediately designed by the ad hoc committee and implemented in the emergency department and the ICU.

Within 60 days, holding time for critical patients in the emergency department was reduced to a minimum of three hours; 24-hour admission delays to the ICU were a thing of the past. The ad hoc committee did its job admirably and then disbanded. This approach worked beautifully for all concerned, most importantly the patients.

ACTION PLAN CHECKLIST FOR REDUCING UTILIZATION COSTS

It takes discipline and commitment for hospital leaders and physicians to work together to reduce clinical utilization costs. When this collaboration works, however, the results can be dramatic. The following steps should be implemented by hospitals seeking to achieve strategic cost reduction success:

- *Step 1: Identify opportunities.* Hospital leaders and physicians should constantly identify opportunities to work more closely together to reduce clinical utilization costs. The ad hoc committee approach is very effective for identifying opportunities and developing potential solutions.
- *Step 2: Use clinical pathways.* Hospital leaders and physicians should work collaboratively to create optimum clinical pathways for treating patients. Clinical pathways can reduce treatment variability and promote efficient use of resources.
- *Step 3: Manage support systems.* Hospital leaders should work diligently to manage systems and resources to support optimal clinical pathways for the benefit of patients and physicians.
- *Step 4: Use education.* Hospital leaders should ensure that physicians understand the full benefits of clinical pathways and how they reduce unnecessary costs through reducing variability of care.
- *Step 5: Initiate corrective actions.* When clinical pathways are not consistently followed, hospital leaders and physicians should work creatively to correct the systemic problems that make it difficult to use pathways efficiently.
- *Step 6: Reward optimal performance.* Physicians who use hospital resources most efficiently should be rewarded. Although this is increasingly difficult from a legal standpoint, hospitals should continue pursuing legal and ethical ways to reward the admirable performance of physicians.
- *Step 7: Address poor performance.* Hospitals should be vigorous in identifying physicians who are not using hospital resources efficiently. All physicians should not be treated alike. Physicians who promote appropriate utilization of resources should be rewarded. Physicians who do not should be encouraged to change and improve their performance. If

this proves impossible, hospitals should encourage those physicians to practice elsewhere.

One remaining category of costs that are directly within the control of hospital leaders is capital costs. Chapter 10 explores the importance of capital costs and opportunities to reduce them through strategic planning and excellent leadership.

Reference

McCaughey, B. 2006. "Saving Lives and the Bottom Line." *Modern Healthcare* (June 30): 23.

CHAPTER 10

Capital Costs

ONE OF THE MOST important factors in the overall cost structure of any hospital is the cost of capital. The cost of capital is profoundly influenced by the overall financial and market performance of the hospital. The best-performing hospitals have access to the lowest-cost capital. Correspondingly, the worst-performing hospitals inevitably pay much higher costs to have access to capital. Some hospitals are so troubled that they may not have access to new capital no matter what the cost may be. These unfortunate hospitals are at risk for closure.

To achieve the lowest cost of capital requires good strategic planning and good operational performance. Hospitals that plan for the future and perform well can have access to the capital they need while achieving the lowest costs of capital. Hospitals that perform poorly rarely, if ever, have effective capital structures. Inevitably, they will have higher capital-related costs. An excellent strategic plan to guide the hospital's future, an excellent capital plan to finance strategic initiatives, and a low-cost structure to achieve the best possible operational performance are common denominators of exceptional hospitals.

IMPORTANCE OF CAPITAL STRUCTURE PLANNING

Highly successful hospitals have both a strategic plan and a strategic capital plan. The strategic plan focuses the hospital on what to do and when to do it. The strategic capital plan focuses on doing the right thing at the right time with financial resources. Hospitals with no long-term strategic plan typically access short-term capital, such as bank financing, lease financing, or working capital financing. These sources of short-term capital are always more expensive than long-term sources, such as bond financing on long-term loans. Preeminent management advisor Peter Drucker once said something to this effect: "Even a bad plan is better than no plan." The same could be said about capital planning for hospitals. Even a less-than-stellar capital plan is probably better and less costly than no capital plan at all. In sum, hospitals with no strategic plan and no strategic capital plan will inevitably pay a higher cost for capital—assuming they can access capital at all.

STRONG PERFORMANCE EQUALS LOWER COSTS

The better a hospital's operating performance, the more likely it will have access to the lowest-cost capital to achieve long-term strategic goals. Hospitals that preform excellently achieve more favorable bond ratings when they access the bond market and correspondingly lower interest rates. Rating agencies use a series of benchmarks to measure a hospital's overall performance. Following are the most important benchmarks (Kaufman 2006):

1. Operating income
2. Excess income
3. Operating EBITDA (earnings before interest, taxes, depreciation, and amortization)

4. Unrestricted cash balance
5. Long-term debt
6. Operating margin
7. Excess margin
8. Operating EBITDA margin
9. Debt service coverage
10. Debt-to-capitalization ratio
11. Cash-to-debt ratio
12. Days cash on hand
13. Average age of plant
14. Revenue collection rate

Of these benchmarks, operating income, debt to capitalization, and days cash on hand are especially important to rating agencies and bond insurers. Additionally, productivity indicators like FTEs per adjusted occupied bed and the ratio of staff and benefit expenses to net revenue are highly scrutinized when bond ratings are established.

More recently, rating agencies and bond insurance companies have begun paying close attention to a hospital's governance performance when issuing ratings and bond insurance. Corporate scandals like Enron, WorldCom, and Tyco have resulted in increased scrutiny of all corporate boards, even nonprofit hospital boards.

Hospitals with negative operating margins, a history of widely fluctuating operating margins, or little or no strategic clarity inevitably pay higher interest rates for access to capital. Lenders perceive these hospitals as higher risks. In turn, they charge higher interest rates to compensate for this additional risk.

ACCESS TO CAPITAL AND STRATEGIC PERFORMANCE

Clarity in strategic direction and comprehensiveness in strategic capital planning ensure that the highest-performing hospitals have access to the lowest-cost capital. These hospitals can build newer facilities and purchase state-of-the-art technology as a result of their excellent strategic performance. Excellent strategic performance leads to better operating performance and, ultimately, access to the best physicians, professional staff, and support staff. In competitive environments, this performance edge undoubtedly results in market dominance.

LEADERSHIP AND CAPITAL COSTS

Providing clarity in strategic direction and capital plans is the hospital leadership's job. CEOs, senior management, and middle management should be committed to thoughtful strategic planning and strategic capital planning. Commitment to developing a well-thought-out capital structure begins with a hospital leadership that understands the value of accessing capital at the lowest possible cost. Management teams with an excellent performance history are able to access capital markets at the lowest possible cost to fulfill their hospital's strategic objectives.

Leadership foresight

A community hospital with a long history of success gradually found itself with declining patient volumes and deteriorating financial performance trends. The CEO, who was a few years from retirement, had the foresight to complete a strategic plan and strategic capital plan that clearly identified the need to build a replacement inpatient campus that would stem patient migration to more modern hospitals 30 miles away. The CEO built a case for reducing hospital costs by 6 percent. This reduction would improve financial performance enough so that the capital needed for the replacement facility would be available at the most attractive interest rate. With the assistance of a strategic cost reduction adviser, he led the hospital through a strategic cost reduction process that centered on empowering the hospital's middle management team.

Strategic cost reduction succeeded beyond all expectations, resulting in an 8 percent green-dollar cost reduction of the hospital's total expenses. Middle managers were the stars of this success story. They recommended eliminating and consolidating four management positions within their own ranks. They used attrition and creative schedule restructuring to reduce staffing by more than 80 employees without a single layoff. They worked with physicians to evaluate supply and medication alternatives, which saved nearly $500,000 annually. Employees submitted ideas ranging from changing oral care kits in the ICU to eliminating several outdated fringe-benefit programs. Physicians responded by voluntarily reducing medical director fees and provided ideas to reduce inpatient lengths of stay by 10 percent.

Two years after this strategic cost reduction initiative, the hospital completed a new, fully insured bond issue with the highest rating and lowest interest rate available at the time. The new inpatient facility was completed two years after the bond issue, immediately reducing the migration of patients to competing hospitals. The CEO, having extended his planned date of retirement by several years, subsequently retired with many accolades from the management team, physicians, and employees whom he admirably had trusted to achieve success in strategic cost reduction.

ACTION PLAN CHECKLIST FOR REDUCING CAPITAL COSTS

Achieving the lowest cost of capital is a necessary element of strategic cost reduction. Key steps to achieving the lowest possible capital costs are as follows:

- *Step 1: Set priority.* The CEO and senior management team need to make achieving the absolute lowest cost of capital one of their leadership priorities. This leadership commitment will ensure that the focus to pursue the necessary strategies will achieve the lowest cost of capital.
- *Step 2: Complete strategic plan.* Hospitals, regardless of size, should have a comprehensive strategic plan. The strategic plan should identify priorities and bring clarity to the hospital's purpose.
- *Step 3: Complete capital plan.* Hospitals should have a strategic capital plan that identifies the sources and uses of capital as well as strategies to access the lowest-cost capital.
- *Step 4: Achieve sound operational performance.* Hospitals that have the lowest cost of capital are those that have the highest level of operational performance. They are the hospitals with the most productive use of labor resources, the most efficient employment of technology, and the highest level of physician and employee morale.
- *Step 5: Reinvest in the hospital.* The highest-performing hospitals that achieve the best operating margins are able to maintain the optimal balance of debt and equity by sufficient investment. They reinvest profits wisely as part of their overall capital structure plan.

This concludes the discussion of opportunities for strategic cost reduction from management costs to capital costs. Part III articulates how to put these opportunities into action by making strategic cost reduction a primary facet of the hospital's leadership priorities.

Reference

Kaufman, K. 2006. *Best Practice Financial Management: Six Key Concepts for Healthcare Leaders,* 3rd ed. Chicago: Health Administration Press.

PART III

Practical Approaches to Strategic Cost Reduction

PART III IS the how-to section of this book. It provides hospital leadership teams with practical approaches for achieving strategic cost reduction success that will work in any type of hospital in any location. Chapter 11 presents concepts on how a hospital best sets the stage for successful strategic cost reduction. Chapter 12 presents my proven method of helping hospitals achieve strategic cost reduction success in 120 days. Chapters 13 and 14 cover alternative approaches that can also lead to strategic cost reduction success by using internal resources exclusively or by using consulting resources to drive strategic cost reduction initiatives. Chapter 15 highlights lessons from failed initiatives and illustrates the pitfalls to avoid when embarking on strategic cost reduction initiatives.

Although Part III is the how-to section, readers should not expect a simple formula for success or an abbreviated checklist that simplifies strategic cost reduction to the status of a to-do list item. Part III provides a framework, and the appendices at the end of the book provide specific examples that can be adapted to any hospital. Strategic cost reduction success comes from leadership commitment and cultural transformation. It is not a project or a program that can be accomplished and then set aside.

Aristotle said, "Excellence, then, is not an act but a habit." To paraphrase his words for a business setting: "Leadership, then, is not a temporary action but a permanent state of mind."

CHAPTER 11

Setting the Stage for Success

ACHIEVING SUCCESS in strategic cost reduction requires the right leaders, the right timing, a well-conceived and well-executed action plan, and excellent communication. Similar to every successful initiative, the right leaders are critically important. Having the right leaders means involving more than a single person, even if that one person is someone as important as the CEO. In other words, achieving success in strategic cost reduction requires the commitment and dedication of board leaders, senior executives, middle management, and medical staff leaders. It is important to note that consultants cannot substitute for a hospital's own leaders.

The right timing is also critically important to success. A hospital that is in panic mode because of financial distress or in a highly distracted state because of multiple challenges or priorities is not a good candidate for strategic cost reduction. On the other hand, a hospital that has a current strategic plan and has thoughtfully reached the conclusion that strategic cost reduction is an opportunity to sustain long-term performance improvements will undoubtedly succeed with a well-conceived and well-executed action plan.

Beyond leadership, timing, and planning, excellent communication is also critical for strategic cost reduction success. It takes comprehensive and focused communication to mobilize not only the hospital's leaders but also its physicians and employees. No strategic cost reduction initiative can succeed without the full knowledge, understanding, and participation of its important physician and employee constituencies.

Finally, success in strategic cost reduction requires selecting the right approach and setting the right goal. Approach and goal-setting concepts will be introduced in this chapter and fully explored in subsequent chapters.

LEADERSHIP COMMITMENT

An insightful and dedicated CEO is essential for strategic cost reduction. Only the CEO can effectively make the case for strategic cost reduction and inspire the internal hospital constituencies to achieve the strategic cost reduction goal. Beyond the CEO, a fully committed senior management team and middle management staff of directors, managers, and supervisors are all essential for achieving a positive strategic cost reduction outcome.

Indeed, successful strategic cost reduction starts with the CEO and senior management, but it does not end there. Leadership commitment from the board and medical staff is also a crucial ingredient for success.

BOARD COMMITMENT

The most highly motivated and effective hospital senior management team cannot achieve strategic cost reduction success unless it has the full commitment and support of the board of directors. The board must understand the need for strategic cost reduction and provide support during the process of change. The process starts with board leaders, especially officers. If the chair, vice chair, and key board committee chairs are supportive of strategic cost reduction, success is much more likely. The board absolutely must view strategic cost reduction as a hospital priority. Once this is apparent, the board must continue to be engaged and informed throughout the process.

Before any strategic cost reduction process, however, the board should review management's action plan, provide suggestions, and formally approve that action plan and strategic cost reduction goal. Board approval helps to ensure that the hospital's key constituencies understand the level of commitment to strategic cost reduction.

Additionally, the board of directors is responsible for any potential conflicts of interest or sacred cow challenges. If the board experiences conflicts of interest that not only are questionable ethically but also increase the overall cost structure for the hospital, the conflicts must be successfully resolved. An example of a conflict of interest that could affect hospital costs would be if a board member's company is selling services to the hospital without the hospital engaging in competitive bidding that would ensure that the lowest costs are being paid. An example of a board-sanctioned sacred cow would be a physician who receives favorable treatment by the hospital because he is an important admitter. This could be a surgeon who refuses to consider alternative sources of orthopedic appliances because of personal preference—a personal preference that forces the hospital to pay 20 percent more for the appliances. Once the board is aware of or has sanctioned the continuation of conflicts and sacred cows, the situations must be addressed so that senior management can make decisions unencumbered by political considerations.

Abuse of power

An academic medical center faced many strategic, governance, financial, and operational problems. The newly appointed CEO decided to begin the renewal process with a new strategic plan and governance overhaul, followed by strategic cost reduction. After launching these initiatives, he immediately ran afoul of a long-time, powerful member of the board of directors. The board member was a prominent citizen of the community, a highly regarded attorney, a large donor to medical center projects, and a member of the CEO's compensation committee.

This same board member was aggressively attempting to sell more legal services to the medical center, personally campaigning to the CEO, in-house counsel, and other members of senior management for more business for her firm. When the CEO resisted, the board member threatened to cut off future donations to the medical center's foundation and suggested, with very little subtlety, that she could negatively influence future CEO compensation decisions.

Fortunately, the medical center's chair of the board intervened. When the offending board member refused to back down, she was asked to leave the board. She did so with a show of great anger, but to the great relief of the medical center's CEO and senior management team.

MEDICAL STAFF COMMITMENT

Members of the medical executive committee, elected and appointed clinical department chairs, and employed medical directors are all critically important hospital leaders in the context of strategic cost reduction. In other words, strategic cost reduction initiatives encompass every aspect of the hospital's operation, including those directly involved with physicians and patient care. Medical staff leaders must be knowledgeable about the case for strategic cost reduction and, ideally, should be active participants in the process to determine which costs can be reduced without harming patient care.

The engagement of the medical staff begins with a thorough explanation of the case for strategic cost reduction. Beyond basic information, medical staff leaders and individual physicians should have opportunities to provide input and suggestions for the strategic cost reduction process as well as opportunities to constructively disagree with proposed cost reduction initiatives. As with the board of directors, there should be a formal review process that allows the medical staff to provide input on cost reduction initiatives before final approval of the strategic cost reduction plan.

KNOWLEDGE BASE

Before beginning a strategic cost reduction initiative, it is important to assemble and review the background trends of key operating departments within the hospital. These trends should include financial and volume trends over, at least, a five-year period. Beyond facts and figures, it is important for the hospital's senior management team to know the status of its middle management, physicians, and employees. Also, any concerns or sensitivities should be identified and addressed before beginning the process. Most hospitals experience heightened anxieties potentially lower morale during a strategic cost reduction initiative. Hospital leaders should be cognizant of these possibilities and should have a specific plan ready to avoid significant declines in physician and employee morale.

STRATEGIC PLAN FOUNDATION

Before any hospital embarks on a strategic cost reduction initiative, it should have a current strategic plan. The absence of a current strategic plan means cost reduction initiatives have no strategic foundation. The strategic plan provides the framework for the all-important case for strategic cost reduction. This case provides the central reason to pursue strategic cost reduction and is the rallying cry for the full participation of all hospital constituencies. The strategic plan and the long-term strategic goals contained therein are absolutely essential in strategic cost reduction.

GOAL SETTING

The hospital's strategic plan provides the foundation for the hospital to set a strategic cost reduction goal. The goal must be expressed in financial terms, usually in the form of green-dollar cost reductions needed to achieve a strategic purpose. For example, a strategic cost reduction goal could be to reduce annual operating costs by $7 million per year for the purpose of improving the hospital's credit rating for a future bond issue. This bond issue could then be used to finance a replacement inpatient tower in 24 months. Following are the key elements of goal setting:

- Establish a specific cost reduction target.
- Set a time period.
- Identify a strategic purpose.

All three elements must be present to meet the definition of a true strategic cost reduction goal.

ORGANIZATIONAL FOCUS

Even when a hospital has the right leadership and appropriate board support, achieving success in strategic cost reduction requires considerable focus. There needs to be sufficient leadership energy and an ability to concentrate. For instance, a hospital preparing for an imminent accreditation survey may not have the time to dedicate to strategic cost reduction. Furthermore, hospitals that are undergoing information system changeovers, comprehensive clinical-quality upgrades, union-organizing campaigns, or other time-consuming initiatives should be very cautious about attempting to implement strategic cost reduction.

Although the challenges of leading modern hospitals make it impossible to find a period of time devoid of all distractions, undertaking strategic cost reduction and other major time-consuming initiatives simultaneously should be avoided.

COMMUNICATION

Successful communication with strategic cost reduction initiatives begins with the CEO. The CEO is the best person to present the case for strategic cost reduction to the rest of the organization. Hospital constituencies need to understand that the CEO not only supports strategic cost reduction but also leads it. Further, the hospital's constituencies need to know that the CEO is committed to implementing strategic cost reduction initiatives proposed by members of management, physicians, and employees.

The CEO must communicate frequently, at least monthly, with members of the board, medical staff, and employees. CEOs who succeed with strategic cost reduction hold town meetings to inform key constituencies about the strategic cost reduction initiatives and provide opportunities for them to ask questions about the process. Letters and e-mails are also effective in keeping constituencies informed. No matter how large the hospital undergoing strategic cost reduction is, it is vital for the CEO to personally communicate with all key constituencies throughout the process.

SELECTING THE RIGHT APPROACH

There are three distinctly different approaches to implementing strategic cost reduction: the internal approach, the use of outside consultants, and a blend of the two. Some hospitals choose an internal approach, which is distinguished by the absence of outside consultants and advisers. The internal approach uses the hospital's own leadership resources to identify and implement strategic cost reduction initiatives. This approach is nearly always the first choice of leadership teams. In some cases, it is completely successful. In other cases, it is a miserable failure.

A second approach is the use of outside consulting resources to plan and implement strategic cost reduction. Large, medium, and small consulting firms are available to work with hospitals to identify and implement strategic cost reduction initiatives. As with the internal approach, sometimes the consulting approach is very successful. At other times, it too is a miserable failure.

The third approach is a blend of internal and consulting approaches. It relies heavily on internal leadership resources while allowing some guidance and technical assistance from consulting advisers. After observing many hospitals that have attempted strategic cost reduction initiatives, I believe the blended approach is the best strategy to fully engage hospital constituencies in the process while ensuring maximum buy-in for sustaining strategic cost reduction success. Regardless of the approach selected, hospitals contemplating strategic cost reduction should determine their readiness before initiating a strategic cost reduction process. Several key readiness questions follow.

KEY READINESS QUESTIONS FOR STRATEGIC COST PREPARATION

Any hospital leadership team contemplating a strategic cost reduction initiative should ask themselves the following questions and be prepared to answer all of them candidly:

1. Is the CEO committed to strategic cost reduction?
2. Are the senior management and middle management teams committed to strategic cost reduction?
3. Is the board committed to supporting the senior management team throughout the process?
4. Does the hospital have a current strategic plan on which to base the case and that will strengthen cost reduction goals?
5. Is the medical staff's leadership committed to working with senior and middle management teams on strategic cost reduction?
6. Does the hospital intimately know its historical trends and performance indicators for the past five years?

7. Can the hospital focus on strategic cost reduction for an adequate period of time (at least 120 days) to achieve success?
8. Is the CEO fully committed to frequent communication throughout the process?
9. Has the senior leadership team decided which strategic cost reduction approach has the best probability of success?

An affirmative yes is desirable for all nine of these readiness questions for the hospital to have the best chance of success with strategic cost reduction. The blended internal and consulting approach is the ideal approach. As hospitals find they cannot achieve the desired outcomes on their own, or have grown disenchanted with the expense and aftermath problems associated with the consulting approach, more of them are turning to the blended approach. The blended approach is what succeeded for Northeast Georgia Medical Center, which was highlighted in Chapter 4, and is the subject of Chapter 12. The internal and consulting approaches are discussed in Chapters 13 and 14, respectively.

CHAPTER 12

120 Days to Strategic Cost Reduction Success

IT TAKES 120 DAYS to transform a hospital's leadership culture to achieve strategic cost reduction success. In less than 120 days, it is nearly impossible to garner the hospital-wide buy-in that strategic cost reduction requires to succeed. And more than 120 days is not necessary if the hospital's leadership is truly committed to cultural transformation. Northeast Georgia Medical Center transformed its culture in 120 days and achieved dramatic success with strategic cost reduction. With an equivalent passion to succeed, any hospital can achieve the same result.

To summarize, strategic cost reduction success requires the hospital's CEO and senior leadership team to be fully committed to cultural transformation. The board of directors must have confidence in its leadership team and fully support transformation efforts. Granted, strategic cost reduction cannot be achieved with consultants alone, but minimal use of experienced advisers can certainly add value to the transformation process. Before the organizational transformation process can truly begin, the hospital must complete a strategic plan for the future. Strategic cost reduction must exist within a strategic framework, which can only be articulated through a strategic planning process that guides the hospital toward achieving long-term goals.

Successful strategic cost reduction requires a leadership approach, not a technical or financial approach. The essence of this approach is a hospital CEO making a case for change, beginning with setting a goal for strategic cost reduction and then empowering senior and middle management to achieve that goal. The CEO must trust senior and middle management. In turn, senior and middle management must trust the input and feedback of their physicians and employees. When senior management is listening to those who know best about where and how to achieve permanent cost reductions and is doing what they recommend, strategic cost reduction is functioning at its peak effectiveness. This is an empowerment process; a cultural transformation process; and, most importantly, a process that really works. Can a hospital truly achieve strategic cost reduction success in four months? The answer is yes. A month-by-month guide with 25 action steps to success follows.

MONTH 1: ADVANCE PLANNING

Month 1 is for planning and getting organized for strategic cost reduction. It begins with careful study of past performance trends and ends with town meetings to prepare all hospital constituencies for active participation in the strategic cost reduction process.

Step to Success 1: Understand Performance History

To plan for strategic cost reduction for future success, a hospital must first understand its recent performance history and current performance status. A thorough review of background trends that highlight key financial and volume trends for the previous five years is the first step. For a suggested list of trends for review, see Appendix A. When a senior management team critically analyzes recent activity and financial trends, it should have a clear picture of the hospital's current performance status and the factors that led to the current performance. Edmund Burke, political philosopher, once said, "Those who don't understand history are destined to repeat it." That sentiment holds true for a hospital that does not understand its own history and past performance.

Step to Success 2: Clear the Decks

Once key trends have been analyzed, it is critical for senior management to "clear the decks" of other projects. It is not possible to embark on a successful endeavor to strategically reduce costs while simultaneously managing a dozen other priorities. Clearing the decks requires the senior management team to have the discipline to temporarily suspend emphasis on secondary challenges and focus primarily on strategic cost reduction and cultural transformation. Clearing the decks also lets everyone know how profoundly important strategic cost reduction is for the hospital's future success.

Step to Success 3: Select an Adviser

The CEO and senior management should select a strategic cost reduction adviser. Options range from small consulting firms to mega-firms that have legions of available consulting resources. Minimal use of consulting resources is advisable when embarking on a strategic cost reduction endeavor; one or two highly experienced advisers and facilitators will be sufficient. An army of briefcase-wielding consultants is excessive and counterproductive. Some hospitals may be tempted to proceed alone, not using any advisory resources. This is usually a mistake. Other hospitals may be tempted to delegate the cost reduction process entirely to an outside consulting firm. This, too, is usually a mistake. Based on my 30 years of hospital executive and advisory experience, the right balance is 80 percent internal leadership and 20 percent outside advisory guidance. Advisers should function as integral members of the leadership team who help the hospital achieve strategic cost reduction success. They should neither lead the effort nor attempt to substitute for the hospital's leadership team.

Step to Success 4: Establish a Calendar

The senior management team should establish a calendar of meetings and progress milestones during the entire 120-day or four-month strategic cost reduction process. This should be done in conjunction with the hospital's adviser and should be detailed enough to note every critical meeting during the four-month process. A sample calendar for a strategic cost reduction process is provided in Appendix B. The calendar provides both discipline and structure to the strategic cost reduction process. Once the calender is set, the hospital's senior management team should never waver from the agreed-on timeline and its key milestones.

Step to Success 5: Make the Case

The case that hospital leadership is making is the rationale for strategic cost reduction. It should be concisely articulated in two or three sentences. It should explain why strategic cost reduction is important in the clearest language possible. It should motivate and inspire everyone in the hospital to participate, and it should create a rallying point for collaboration throughout the hospital so that the strategic cost reduction goal is reached. For an example of a well-articulated case, see Appendix C.

Step to Success 6: Set the Goal

The strategic cost reduction goal should be expressed in terms of annual green-dollar cost reductions. The goal should be set high enough to ensure that the hospital is able to make its case and meet its strategic performance goal. The goal should be set by the CEO and senior management in conjunction with the hospital's strategic cost reduction adviser and with input as necessary from interested third parties, such as the hospital's strategic planners, bond insurers, lenders, or rating agencies.

To begin the process of strategic cost reduction, the hospital needs to set a specific financial performance target. Once the performance target has been identified, it is recommended that the target is adjusted by approximately 20 percent to allow for contingencies. For example, if the performance target to achieve future strategic goals is determined to be $5 million in cost reductions annually, it is recommended that the actual strategic cost reduction goal be set at $6 million.

Step to Success 7: Prepare the Board of Directors

The hospital's board of directors must be fully apprised of the need for strategic cost reduction and the challenges the organization will face during and after strategic cost reduction is implemented. The board of directors not only needs to fully support the process, but it also needs to be prepared to support the hospital's CEO and senior management team during the process. There will be ample opportunity for second-guessing both inside and outside the hospital. The board of directors must never waver in its support; otherwise, the second-guessers and naysayers could prevail in derailing strategic cost reduction success.

Step to Success 8: Organize Task Forces

Strategic cost reduction task forces should include all members of senior and middle management. Although titles vary greatly from hospital to hospital, this group would include the directors and managers from all departments and nursing units throughout the hospital. In small hospitals, this group may be composed of 30 to 40 members. In larger hospitals, it may exceed 100 members. Regardless of titles, all members of senior and middle management who directly control hospital costs must be included on the task forces. Once all members are selected, the next challenge is to divide the members into working task forces. A task force with approximately seven to ten members is ideal. In small hospitals, a total of five or six task forces is adequate, whereas a large hospital may need ten or more task forces to achieve the optimal task-force size. Senior management, which includes those at the vice president level and above, should be organized into its own task force. Remaining task force members should be divided according to the following useful guidelines.

First, task forces should be heterogeneous. Membership of each task force should be a diverse mixture of clinical and nonclinical managers and directors. Ideally, a mixture of clinical, nursing, business, support services, and technical services should be represented on each task force. In addition, task force membership should be varied in terms of hospital tenure, age, and experience. The more heterogeneous the task forces, the better the strategic cost reduction outcome.

Step to Success 9: Plan a Kickoff Retreat

Once task forces have been selected and organized by senior management, a kickoff retreat should be planned. The CEO and senior management should be the primary speakers at the kickoff retreat. Here, the case for strategic cost reduction should be articulated, and the CEO should specifically express his trust and confidence that the hospital's management team can achieve the strategic cost reduction goal. It should be emphasized that advisory resources will be used sparingly and that the CEO is relying on members of the management team, with input from physicians and employees, to develop the cost reduction initiatives necessary to achieve the strategic cost reduction goal. The adviser's role in the kickoff retreat should be to clarify the concept of green dollars, provide examples of strategic cost reduction success with previous client hospitals, and act as a resource to answer questions about the task force process and methodology for identifying cost reduction initiatives.

Step to Success 10: Conduct Town Meetings

After the kickoff, the CEO should conduct hospital-wide town meetings for all physicians and employees. At these meetings, the CEO should present the case for strategic cost reduction, the financial goal, and the task force methodology to be used when achieving the goal. The emphasis should be on making it clear that all members of management, physicians, and employees will be working as a team during the process. Plenty of time should be allowed for the CEO to answer questions. See Appendix D for a sample of questions physicians and employees typically ask during this early stage of a strategic cost reduction effort. At the end of month 1, the hospital should be fully prepared to proceed. Months 2 and 3 are challenging and intense but immensely rewarding in terms of organizational development, cultural transformation, and identification of green-dollar cost reduction options.

MONTH 2: GETTING ORGANIZED FOR SUCCESS

Month 2 is when the hospital's task forces, with the help of physicians and employees, go about identifying initiatives that can permanently reduce green-dollar hospital costs. Month 2 provides an opportunity for collaboration and teamwork throughout the hospital. As with month 1, month 2 ends with town meetings to keep all hospital constituencies advised of strategic cost reduction progress.

Step to Success 11: Assign Task Force Goals

At the beginning of month 2, each task force should be assigned a specific financial goal. The goal should be apportioned according to the percentage of the specific operating costs that are controlled by members of the task force. In other words, if members of task force 1 collectively control 10 percent of the hospital's costs, they will be apportioned 10 percent of the strategic cost reduction goal. This is a fair and equitable way to distribute goals for the various task forces. Note that individual task force members are not given a financial goal; this goal is a team goal. Finally, no across-the-board cost reductions should be assigned as part of the strategic cost reduction process.

Step to Success 12: Provide Task Force Training

Soon after the initial kickoff of strategic cost reductions, dedicated training of task force members should proceed. This should be accomplished by the adviser in conjunction with the CEO. The training objective is to articulate a series of guidelines for the task forces to follow to achieve the intended successful result. Although each hospital is unique, Appendix E provides a list of task force guidelines that have been used very successfully by small, medium, and large hospitals. Part of the initial focus during month 2 is ensuring that each task force member has detailed information about his department's historical trends and financial indicators. To help with this, it is recommended that a member of the human resources department and a member of the finance department be assigned to each task force as task force support staff. This will give each task force immediate access to current and historical information on staffing from the human resources department and data on supplies and expenses from the finance department.

Step to Success 13: Select Task Force Chairs

Also in the beginning of month 2, each task force should elect a chair and a co-chair. It is important for the chair and co-chair to receive additional training from senior management and the hospital's adviser on the process of facilitating teamwork meetings. A set of overall guidelines about how task force meetings should be run is provided in Appendix F. It is also important that each task force establish a separate calendar for meetings during the 60-day idea-development time frame that takes place in months 2 and 3. At minimum, each task force should meet two to three times per week for at least 90 minutes. These meetings are for brainstorming and idea-sharing. The first few meetings should be dedicated to allowing each member of the task force to describe, in detail, her department's labor, supply, service, and clinical utilization costs.

The leadership role of the task force chair and co-chair is extremely important at these early meetings. Naysayers tend to express themselves with phrases like "We are already running too lean," "There is simply not another dollar that can be saved in my department," and "There are plenty of savings to be achieved in department X, because they never do their fair share." If the CEO has done his job properly during the kickoff retreat, there will be little ambiguity in the expectation that every member of every task force is to participate in the strategic cost reduction process.

Step to Success 14: Begin Meetings with Strategic Adviser

After the task forces are organized and have received their initial orientation and training sessions, the next step is to begin meeting with the hospital's strategic cost reduction adviser. This adviser should meet with each task force for approximately 90 minutes at least every 10 days during months 2 and 3. The adviser should be a sounding board and should provide ideas from previously successful strategic cost reduction situations. The adviser can also be a facilitator, coach, and critic when necessary. The best strategic cost reduction adviser is one whose experience allows him to thoughtfully evaluate proposed cost reduction ideas from task force members and to help the proposing director or manager refine those ideas. Furthermore, it is helpful if the adviser is experienced enough to help directors and managers understand the difference between hypothetical savings and true green-dollar savings. The acid test for whether an idea represents a true green-dollar savings is to see whether the facility is able to discontinue writing checks or making electronic funds transfers after the implementation date of the proposed initiative.

Step to Success 15: Initiate Weekly Progress Reports

During month 2, the process of making weekly reports of the task force's activity to the CEO and senior management team should be initiated. A sample weekly report from a task force is provided in Appendix G. Weekly reports provide the CEO and senior management with a status report of the task forces' activities and their initial cost reduction ideas. At an early stage, the reports allow senior management to provide positive and negative feedback to task force members. The weekly reports also provide the CEO and senior management with insights on how effective individual directors and managers are in using the task force approach.

Step to Success 16: Make a Decision on Staff Reduction

During month 2, senior management will be faced with a critical decision; task force members will want to know whether or not they will be expected to make labor-force reductions. In most cases, strategic cost reduction goals cannot be met without certain reductions in labor costs. It is up to senior management to determine whether or not the task forces will be charged with using employee turnover, such as resignations and retirements, to meet labor-force reduction goals, or whether it will be necessary to use such means as a layoff to reduce the labor force. If a reduction in labor force is determined to be necessary, each task force should be fully informed of this expectation and assigned to make recommendations for the most appropriate areas for labor reductions that would not compromise patient care quality or customer service.

Step to Success 17: Repeat Town Meetings

Near the end of month 2, it is advisable for the CEO to hold another series of town meetings for the purpose of updating physicians and employees on progress of the strategic cost reduction and, more importantly, answering questions or commenting on any rumors that may have surfaced since the initial town meetings.

MONTH 3: REFINING COST REDUCTION INITIATIVES

Month 3 is for refining cost reduction ideas and making necessary commitments to overcome resistance to the changes the cost reduction ideas represent. During month 3, all task forces should have accumulated sufficient cost reduction ideas to meet their assigned goals.

Step to Success 18: Solicit Ideas

At the beginning of month 3, each task force should have a solid list of proposed cost reduction ideas for staffing, supply, service, and clinical utilization. These ideas should be based on the task force members' best judgments about where expense cuts can be made without harming quality or service in their respective departments. Each task force member should be expected to talk candidly with physicians and employees in those departments about proposed changes. This provides a vehicle for the widespread participation of all task force members involved in strategic cost reduction brainstorming.

Step to Success 19: Refine Ideas

The two- and three-times per week meetings with the task forces should be used to refine the strategic cost reduction ideas. The meetings with the outside adviser that take place every ten days should be used to further refine the ideas and begin the process of planning for implementation. By the middle of month 3, each task force should be at or near its assigned strategic cost reduction goal. Each task force member should be expected to have contributed green-dollar ideas to the goal by the middle of month 3. Any director or manager who has not participated at this point should be counseled by their administrative supervisor. Although individual goals are not assigned to individual task force members, it is expected that everyone will offer green-dollar cost reduction ideas for their department. Failure to participate should be grounds for the hospital to consider firing the recalcitrant director or manager.

Step to Success 20: Overcome Resistance to Change

Experience has proven that during month 3 pockets of resistance may remain among reluctant directors and managers. These individuals may absolutely refuse to participate or propose ludicrous ideas that they know will be rejected. This kind of passive-aggressive behavior is a clear indication that the individual is neither willing nor able to participate in the cultural transformation process so vital to the achievement of strategic cost reduction goals.

On the other hand, task forces that are at or near their assigned goals will likely be excited about their progress and proud of their proposed ideas; a task force goal that is achieved through hard work, creativity, and empowerment creates a tremendous sense of enthusiasm. Experience has demonstrated that if the hospital's leadership does its job during the advance planning of month 1, their management staff will succeed during months 2 and 3.

MONTH 4: MAKING FINAL DECISIONS AND PLANNING IMPLEMENTATION

Month 4 is for making decisions, planning for implementation of cost reduction ideas, and celebrating success. At the end of month 4, as with previous months, town meetings should be held to share the hospital's strategic cost reduction success story with all physicians and employees.

Step to Success 21: Say Yes

By the beginning of month 4, the CEO and senior management should have reviewed the weekly task force reports thoroughly and evaluated the proposed cost reduction ideas for potential implementation. As a general rule of thumb, it is senior management's job to say yes. If the director or manager believes she can implement the proposed cost reduction idea, senior management should rarely disagree. In fact, if they say no too often, senior management may lose credibility with its own middle management staff. This can be a real test for senior management. For instance, directors and managers may propose changes that affect the hospital's sacred cows. Sometimes, these sacred cows are prominent physicians and board members with conflicts of interest. This is an opportunity for administration to show leadership strength by looking for ways to approve the initiatives that may negatively affect sacred cows, rather than ignoring or discounting them.

Step to Success 22: Plan Necessary Staff Reduction

At the beginning of month 4, it should also be apparent to senior management whether or not a reduction in the labor force will be necessary as part of the strategic cost reduction process. If it is necessary, labor counsel should be retained at this time to begin the planning process for the labor reduction. Appropriate legal reviews and advance planning must be accomplished during month 4 if a staff reduction is needed to achieve the strategic cost reduction goal.

Step to Success 23: Make Final Decisions and Create Board Report

By the middle of month 4, senior management should be in a position to make final decisions about all proposed strategic cost reduction initiatives from all task forces. Ideally, each task force should meet or exceed its assigned goal. However, what is most important at this point is that the cumulative proposed strategic cost reduction initiatives add up to the goal. It is not necessary for every task force to achieve its goal if the overall hospital cost reduction goal is achieved. When senior management has made final decisions about each of the task force's recommendations, the decisions should be assembled into an overall report for the board of directors. The overall report should be a synopsis of the savings initiatives. A sample of an overall report is provided in Appendix H. Additionally, a detailed listing of each cost reduction initiative by department should be assembled. A sample page of a detailed cost reduction report is provided in Appendix I. At the end of month 4, a leadership celebration retreat should be scheduled that includes the board of directors, medical staff leaders, and all members of management who participated on the task forces.

Step to Success 24: Plan a Celebration Retreat

Each task force should be asked to make a presentation at the celebration retreat. Presentations should include samples of cost reduction initiatives, learning experiences throughout the process, and additional suggestions for future improvement in the hospital's customer service and communication practices. The final retreat gives the hospital's middle management a prominent opportunity to present their work to the board of directors and medical staff leaders. The leadership retreat can be an extraordinarily uplifting celebration that gives the board of directors an opportunity to recognize and give thanks for the hard work and dedication of their senior and middle management. The value of this positive reinforcement is priceless.

Step to Success 25: Seek Formal Board Approval and Conduct Town Meetings

After the celebration retreat, the board of directors should formally adopt the strategic cost reduction plan. Formal approval is the initial step in actually implementing the plan. After the board's approval, a third series of town meetings should be scheduled for medical staff members and employees. This final town meeting should formally give credit to the dedication of the task forces and enthusiastically present the plans for implementing the strategic cost reduction initiatives.

KEYS TO FUTURE SUCCESS

The 25 steps to strategic cost reduction success and the cultural transformation that makes success possible can take as few as 120 focused and efficiently run days. Implementation and follow-up monitoring must be continuous. Several critical success factors determine whether a hospital will succeed not only in developing a strategic cost reduction plan but also in implementing and sustaining it over time.

The initial success factor is the high quality of the hospital's leadership. Clearly, the quality of senior management and middle management is rigorously tested during a strategic cost reduction process. Only the strongest members who have contributed to the strategic cost reduction plan are going to be able to implement and sustain the initiatives in the future. If the initiatives cannot be sustained, then decisions about the members of management responsible will have to be made. It is unfortunate, but true, that some managers, even if they have the best attitudes and intentions, simply cannot sustain the strategic cost reduction initiatives that they themselves proposed. Unless there are extenuating circumstances, those people should be considered for reassignment or termination. Harsh as that may sound, if the strategic cost reduction initiatives cannot be sustained, then the hospital's cultural transformation will be incomplete and temporary.

Leadership development and team building are other future success factors. Collaboration and teamwork skills will be honed during a successful strategic cost reduction process. Those skills must be continuously updated for the true value of the cultural transition to be visible. The value of frequent and candid communication that takes places place during the four-month process should be continued and expanded. The use of town meetings for the CEO and other members of senior management to take questions and answer them candidly is especially important.

The task force methodology can and should be used for future challenges. Hospitals that have successfully used the task force methodology with strategic cost reduction have found that the same process can be used for planning new facilities, improving customer service, creating or improving clinical pathways, and facing virtually any other hospital-wide challenge that requires collaboration and teamwork.

Finally, one of the most important critical success factors is the frequent and continuous communication among the members of senior management, directors, and managers who do succeed in implementing strategic cost reductions and sustaining the results over time. There is no substitute for achieving a challenging goal. Sustaining goal achievement over time is cause for celebration; the CEO and senior management should take every opportunity to reinforce success and give due credit to physicians, employees, and everyone who contributed.

Although the blended approach of internal leadership resources and limited external advisement described in this chapter is most highly recommended for strategic cost reduction success, some hospitals choose to use an approach that relies solely on internal leadership. The challenges of this approach are discussed in Chapter 13.

CHAPTER 13

Internal Approach

GOOD HOSPITAL LEADERSHIP teams believe they have the solutions to most challenges. This may be especially true with such challenges as strategic cost reduction. When faced with less-than-optimal financial performance, hospital leadership teams naturally gravitate toward fixing the problem themselves. It is safe to say that most hospital boards of directors are comfortable with this approach. Boards naturally presume that their leadership teams have the skill and talent to achieve whatever strategic cost reduction goals are necessary for the long-term success of the hospital.

This do-it-yourself mentality may also be present with lenders, rating agencies, and debt insurers; they collectively assume hospital leadership teams will achieve the desired levels of performance. When this does not happen, they expect management to take appropriate action. In an extreme case, such as lending covenant defaults, retention of consultants is usually automatic. Short of covenant defaults, the use of outside consultants is often resisted by hospital leadership teams and their boards. "We don't need the help," "it costs too much," and "it takes too long" are common refrains.

Because of those problems, and many others, the internal approach tends to be selected first by hospital leadership teams contemplating strategic cost reduction initiatives. In some cases, the do-it-yourself approach can be highly effective. Success in the internal approach begins by setting the cost reduction goal.

STRATEGIC COST REDUCTION GOAL

The strategic cost reduction process begins with selecting the financial performance improvement goal. Historical performance is often a useful place to begin. Although the hospital may not be performing optimally at present, most hospitals have periods in their recent history (five years) when financial performance was at least adequate. A review of the benchmarks for the adequate performance years should take place before establishing strategic financial performance improvement goals for the future. In other words, pick a good recent year, and study it carefully to learn which performance parameters were in place. Next, determine whether those past successful performance parameters can be duplicated in future years.

It is advisable to compare current financial performance parameters with the prior year's positive benchmarks. Where are the key differences? Was productivity higher? Were supply expenses lower? Were clinical utilization patterns more closely aligned with case-mix indicators? By reviewing answers to each question, a hospital leadership team can learn from past success factors. Also, reviewing the rating agency performance parameters and benchmarks from peer groups, such as state hospital associations and voluntary hospital association groups, can be very useful in determining benchmarks for future success. With the help of these sources, the hospital's leadership team can set the strategic cost reduction goal.

GETTING STARTED

Once the financial performance improvement goal is established, the hospital needs to organize the process of strategic cost reduction. When the internal approach is used, the financial performance improvement goal is typically divided among the hospital's senior management according to the specific budget for each senior management member. For example, if the vice president of patient care services controls 50 percent of the total operating budget for the hospital, that vice president will be responsible for achieving 50 percent of the strategic cost reduction goal. In other words, the hospital's senior management team becomes personally accountable for achieving a portion of the overall performance improvement goal based on the operating budget senior management directly controls.

A key question that needs to be answered by each hospital at this stage is whether or not the cost reduction goal will be achieved through labor-force reductions. If reductions are going to be considered, the next question is whether or not labor-cost reductions will be made through layoffs or through attrition. Once these questions are answered, the specific activities to reduce labor and other costs can begin.

Again and again

A small rural hospital struggled for more than a decade with improving financial performance. Each year, the CEO and administrative staff prepared a budget for the next financial year, and every year the draft budget showed a loss for the coming year. In turn, this triggered a frenzy of budget-cutting activity led by the CFO.

The CFO dutifully met with all department directors and gave them budget cuts for the next year. The CFO incorporated these budget cuts into the financial budget prepared for the board of directors' approval. No follow-up ever took place, and, as a result, no green-dollar cost reductions were ever implemented; the department directors knew they would not be held accountable.

Even though the budget balanced every year on paper when presented to the board of directors for approval, it never balanced in reality. This situation is reminiscent of Albert Einstein's definition of insanity, "doing the same thing over and over and expecting a different result." After ten years of financial deficits, the hospital exhausted all of its cash reserves. The CEO was fired, and a strategic planning and strategic cost reduction process was immediately implemented by the successor. The next year the hospital posted its first profit in a decade.

COST REDUCTION ASSIGNMENTS

When a hospital is using the internal approach, cost reduction assignments are usually expressed as a percentage of an individual's budget. For example, if the overall goal is $6 million in cost reductions, and this goal represents 5 percent of the overall hospital's budget, the 5 percent parameter can be used to make specific assignments to specific members of management. Typically, each member of middle management will be assigned a goal based on a percentage of his department budget. If the overall goal is a 5 percent reduction, each department head, nurse manager, or director is given an assignment to reduce costs by 5 percent. Members of middle management are then challenged to come back to their administrative supervisor with cost reduction ideas to achieve those 5 percent reductions.

POSITIVE ATTRIBUTES

The most positive attribute of the internal approach is flexibility. The internal approach can be used at any time, depending on the hospital's needs. Strategic cost reductions can be organized quickly or slowly, and the time frame can be short or long, depending on the urgency of the hospital's needs for financial performance improvement. The internal approach avoids the additional time and expense of using outside consultants.

The internal approach can impart a sense of ownership by members of the hospital's leadership team; they can be justifiably proud that they achieved a positive cost reduction outcome themselves, rather than depending on outside resources. Finally, use of the internal approach can produce a more positive response to strategic cost initiatives among the various hospital constituencies. In other words, a hospital that succeeds in internally identifying strategic cost reduction initiatives may have a higher probability of implementing and sustaining those reductions because it was responsible for identifying the changes in the first place.

NEGATIVE ATTRIBUTES

The internal approach does have its limitations. It lacks constructive challenges that an outside adviser would contribute when identifying strategic cost reduction ideas. Depending solely on internally developed cost reduction ideas can limit the scope of available ideas and potentially hinder the ability to make changes. The internal approach is subject to complacency and failure of the entire strategic cost reduction process. If the only time a hospital thinks about cost reduction is at budget time, the organization is unlikely to manage its costs successfully throughout the year. Indeed, budget time is undoubtedly the most common time for the internal approach to be used. Cost cutting with no real strategic foundation is a short-term approach at best. Failing to produce a bottom line that is positive enough for next year's budget can motivate many hospital leadership teams to cut the budget to meet the following year's financial performance target. This is not strategic cost reduction. A more accurate label would be budget-balancing cost reduction.

BOTTOM-LINE CRITIQUE

The internal approach to strategic cost reduction should always be considered first. However, it should not be considered in the absence of alternatives. The internal approach can be quickly instituted and, in some instances, can be extraordinarily successful. However, in most instances, the internal approach is simply not rigorous enough to achieve and sustain long-term strategic cost reduction goals. Recurring failure of the internal approach can compromise a hospital leadership team's credibility in the eyes of its board of directors, especially if budgets with positive bottom lines are presented to the board year after year for approval and are never met. When the internal approach is not enough, some hospitals turn to consultants for answers.

CHAPTER 14

Consultant Approach

CONSULTANTS CAN GIVE valuable assistance to a hospital seeking to achieve strategic cost reduction goals. Consultants can supplement the hospital's internal expertise in specific areas such as staffing, fringe benefits, supply expenses, and clinical utilization. The synergy added by consulting expertise can sometimes achieve better results than relying on internal leadership resources alone to achieve cost reduction goals.

If the consulting approach is used, selecting and managing the best consulting company can be challenging. If the hospital is committed to selecting the absolute best consulting company and devoting the energies necessary to actively manage the consulting engagement, the results can be positive. A word of caution: A mere decision to use consultants for strategic cost reduction in no way ensures success. The deciding factor of achieving strategic cost reduction success still rests with the hospital's leadership.

SELECTING THE RIGHT CONSULTANT

Consulting company options for hospital cost reduction abound. However, not all consulting companies are appropriate when considering the implementation of strategic cost reduction. Some consulting resources are better suited for turnarounds and short-term expense reduction. A hospital considering strategic cost reduction must have a longer-term outlook and select consultants with the greatest correlation to that longer-term outlook. If a hospital needs a financial turnaround, then by all means it should hire turnaround consultants. Turnarounds must not to be confused with strategic cost reductions, however. In a turnaround, the hospital's survival is at stake. Profound changes may have to be made in the interest of short-term survival, including cutting costs to the extent that patient care quality and customer service may suffer. With strategic cost reduction, survival is not the challenge. Rather, the hospital is seeking to perform better to achieve its strategic priorities. Strategic cost reduction supports the accomplishment of strategic goals; reducing patient care and customer service quality is not an option.

There are large, small, and a myriad of medium-sized consulting companies. There are specialized consulting companies that focus on revenue enhancement, cost cutting, reimbursement maximization, and other financial disciplines. Some consulting companies promote themselves as leadership advisers, but others take pride in their technical expertise.

A hospital must consider its specific needs before embarking on the consultant selection process. Once it has decided which type of consulting company would provide the greatest possible assistance, the next challenge is deciding which specific firm has the most compelling skills, the best chemistry with the leadership, and the best compatibility with the hospital's value system. A well-organized request for proposal, reference checks, and personal interviews are all important prerequisites to selecting the best consultant. When the selection process is complete, managing the consulting company is the next challenge.

MANAGING CONSULTANTS

If consultants are selected to help the hospital with strategic cost reduction, due consideration should be given to introducing, communicating, and coordinating the consulting assignment so that the highest probability of success is achieved. Introducing the consulting company to hospital constituencies in a positive way is a must. Ideally, the CEO introduces the consulting resource via written communication and personal introductions during the appropriate board and management meetings. The CEO should take great care in assuring the hospital's constituencies that the consultants are there to help achieve cost reduction goals but not to substitute for the hospital's leadership team.

Frequent communication with all constituencies and the consulting company is vital throughout the consultant-assisted strategic cost reduction process. Maintaining control of the consulting engagement is the responsibility of the CEO and senior management. At no time should it appear to any hospital constituency that the consultants are in charge. At the beginning of the consulting engagement, senior management should organize a calendar with specific milestones. Ideally, the calendar would include the dates, times, and locations of all key meetings. The CEO and senior management team should meet with the consultants throughout the engagement so that face-to-face communication is frequent and thorough. The consultants should focus their energies on being a positive force throughout the hospital, rather than a negative or destructive force.

> **Feeding frenzy**
>
> An urban hospital with many financial challenges hired multiple consulting companies to help it improve financial performance. It hired one consulting company to do revenue cycle work. It hired a second company to perform supply chain miracles. It hired a third company to apply staffing benchmarks to all departments and nursing units to reduce labor costs. And it hired a fourth consulting company to seek ways to cut employee fringe-benefit costs.
>
> After more than a year with the four consulting companies, the hospital had spent more than $10 million on consulting fees and

had not materially improved its financial performance. This failed approach finally came to an end when a long-time clinical department director wrote to the chair of the board, "Our hospital has become a feeding frenzy for consultants. There are so many consultants, they trip all over each other and contradict each other's ideas. I can't get anything done because I spend all day answering consultant's questions. My staff thinks we have all gone mad and that consultants have completely taken over. When will it end?" It ended shortly thereafter with the removal of the CEO and all four consulting firms.

ORGANIZATION OF THE PROCESS

Hospital leaders and consultants should jointly create a teamwork approach for strategic cost reduction. Ideally, the team will be cohesive and make the best use of each other's strengths. Consultants retained to help the hospital with strategic cost reductions often rely heavily on benchmarks. Care should be taken so that the benchmarks selected are as relevant and as current as possible. Experience teaches that members of middle management will be highly resistant to the use of benchmarks that they do not see as pertinent to their hospital.

A system of progress reporting should be set up so that consultants and hospital senior management are well informed about strategic cost reduction initiatives. Frequent communication with other key constituencies should be scheduled. During this communication, care must be taken to ensure that the consultants are never dominating.

BENCHMARKS AND ASSIGNMENTS

When the consulting approach is used for strategic cost reduction, applying benchmarks can provide a guideline for identifying opportunities for improvement. In all hospitals, some departments and services are better managed than others. Some departments demonstrate higher labor productivity than others; the use of benchmarks can constructively point out differences and opportunities for improvement. Consultants, meeting with their assigned hospital management counterparts, can and should create an environment of creativity and objectivity. Consultants should be experienced enough to contribute their own ideas and provide advice and guidance on implementing strategic cost reduction initiatives.

POSITIVE ATTRIBUTES

One of the most beneficial aspects of using the consultant approach is that new ideas may surface because experienced outside advisers are part of the strategic cost reduction process. Ideally, consultants working with hospital leaders have experience in helping other hospitals succeed with strategic cost reduction and can plant the seeds of some of their previous clients' successes into their current client's establishment. Also, consultants can question and challenge the status quo of the organization. Most hospitals have some built-in resistance to change; changes to how they spend money also meets resistance. The consulting approach can constructively encourage hospitals to rethink their green-dollar costs.

Consultants can stimulate creativity and develop alternatives to staffing, supply, and service costs. The wide range of experiences that consultants bring to their clients can be of great benefit. The use of the consulting approach can add discipline and structure to the strategic cost reduction process. Ultimately, when a hospital is paying, sometimes handsomely, for outside advice, it is inclined to pay attention and to do something constructive with that advice.

NEGATIVE ATTRIBUTES

The use of consultants for strategic cost reduction has certain risks. Some members of the hospital staff may perceive that consultants have taken over. Leadership of the hospital must not be abdicated, especially during a strategic cost reduction process; it should never appear that consultants are leading the process. That job remains in the hands of senior management.

Consultants are expensive. It seems ironic to some hospitals that they would need to spend money to save money. When considering the value of consulting dollars, the return on investment becomes critical. If a hospital spends $1 million on cost reduction consultants and only manages to find $500,000 in green-dollar savings, obviously the investment was not a wise one. In that unfortunate scenario, the next consultant the hospital retains could very possibly be a search consultant for a new CEO.

Consultants can also be distracting. This is especially true if the consulting company uses a large number of consultants (more than five) on the strategic cost reduction engagement. No matter how professional, consultants can divert the hospital's leadership energy and attention from their principal responsibility of leading the hospital.

Another potential risk with using consultants for strategic cost reduction is the lack of sustainability of cost reduction initiatives. If hospital physicians and employees perceive that all of the cost reduction ideas were proposed by consultants, the long-term survival of those ideas after the consultants have departed is highly questionable.

CRITIQUE

Certainly, there are good consultants and bad consultants. Also, there are good consulting approaches and bad consulting approaches. The main critique of the consulting approach when applied to hospital strategic cost reduction is that consultants can never substitute for good leadership. Although consultants can sometimes contribute excellent ideas and stimulate creativity, it is still the responsibility of the hospital's senior and middle management teams to implement and sustain long-term cost savings so that strategic cost reduction goals may be achieved. Consultants cannot do this, no matter how talented they may be. The hospital's CEO and senior management team must always be aware of this limitation and avoid the temptation to delegate this important task to outside advisers.

The blended approach, internal approach, and consulting approach all have their advocates. Regardless of the approach hospitals use to strive for strategic cost reduction success, certain pitfalls should be anticipated and avoided. A description of these pitfalls follows in Chapter 15.

CHAPTER 15

Pitfalls to Avoid

WHEN A HOSPITAL seeks to strategically reduce costs, much is at stake. Success can make the hospital stronger, more competitive, and better able to meet the needs of its community and its patients. Failure, on the other hand, can compromise the hospital's ability to meet its obligations to the physicians, employees, patients, and community. As the saying goes, "Failure is not an option."

Many pitfalls should be avoided when a hospital embraces the concept of strategic cost reduction. Failure to plan strategically is one of the most common mistakes. In fact, this failure dooms strategic cost reduction before the hospital even begins the process. No matter what goal the hospital sets for itself to strategically lower its costs, it needs to succeed. Failure to achieve the goal will have multiple effects, all of them negative. The hospital that fails at strategic cost reduction loses credibility, both internally and externally. Failure can mean increased problems with quality, morale, and customer service. Ultimately, failure to implement strategic cost reduction initiatives, or failure to carefully monitor initiatives after implementation, diminishes the positive impact of strategic cost reduction. These are just some of the pitfalls that can compromise the success of a strategic cost reduction initiative.

Undoubtedly, the leadership team that fails to achieve its strategic cost reduction goal puts their hospital and their personal careers at risk. Knowing the potential pitfalls and addressing them prospectively can make the difference between success and failure.

FAILURE TO PLAN

Hospitals that fail to achieve strategic cost reduction success oftentimes fail because they do not invest the time in advance to create a strategic plan for the hospital's future. The strategic plan provides the basis for the strategic cost reduction case and the framework to create the strategic cost reduction goal. There is no alternative for starting with a good strategic plan. Failure to do so does not necessarily mean that attempts to lower costs will completely fail. However, without the strategic planning foundation, the likelihood of long-term strategic cost reduction success is greatly diminished.

PARTIAL GOAL ACHIEVEMENT

A hospital that does not entirely achieve its strategic cost reduction goal is subject to a loss in confidence. For example, if a goal of $10 million in strategic cost reductions is embraced and only $6 million in cost reductions are achieved, the shortfall cannot be hidden. Additional cost cuts will be necessary to make up the difference. In turn, the board, medical staff, and employees may lose faith in the hospital's leaders. This loss may also be felt among external parties such as lenders, rating agencies, and bond insurers.

The hospital that fails to meet its strategic cost reduction goals may be forced to retain consultants to identify additional cost reductions. In the instance where the hospital was using the internal approach, the subsequent use of consultants may be viewed as a leadership failure. If the hospital had already used consultants and still failed to achieve strategic cost reduction goals, the use of additional consultants will assuredly be seen as a failure in the eyes of physicians and employees, and probably the board as well.

INCREASED OPERATIONAL PROBLEMS

Operational problems may emerge if strategic cost reduction goals are not met. Quality is likely to suffer during the day-to-day operations of the hospital. If additional cost reductions are not forthcoming to meet the original strategic cost reduction goals, it is inevitable that customer service and clinical quality problems will develop. Cutting corners to meet strategic cost reduction goals is common. Unfortunately, those reactionary cost-cutting measures are often inappropriately deployed. For example, it is not uncommon for leadership teams that are struggling to control labor costs to enforce overtime eliminations, hiring freezes, or other temporary measures. Inevitably, these measures fail to achieve the desired green-dollar results. Physicians and employees will become concerned about the future, both for themselves and for their hospital, and increased turnover, attitude problems, and service declines will become commonplace.

If financial goals are not being achieved, the response of some senior management teams is to implement budget cuts throughout the hospital. Edicts from the corner office to reduce costs by some percentage to meet the budget are commonplace under such circumstances. The annual budgeting cycle becomes a nightmare of repeated budget slashings to meet budget-reduction goals. Under these circumstances, strategic cost cutting might more appropriately be labeled as reactionary cost cutting.

FAILURE TO IMPLEMENT

Another pitfall that is sometimes experienced by otherwise successful hospitals is the failure to fully implement strategic cost reduction initiatives. Sometimes excellent ideas surface during the strategic cost reduction process, but they never become fully or effectively implemented. What sounded like a good idea on paper has no effect in the hospital unless the changes are actually implemented successfully. In other instances, strategic cost reductions may have been implemented, but the desired results are not forthcoming. Sometimes strategic cost reduction initiatives have unintended consequences that the hospital did not fully anticipate. In some instances, these hospitals may actually experience cost increases rather than the intended cost reductions.

If the hospital does not achieve its strategic cost reduction goals, senior management may find themselves in a new and less flattering light. Middle management and employees will certainly point the finger of blame at senior management when strategic cost reduction goals are not met. If senior management then turns to outside consultants to remedy the situation, further losses of credibility are likely. The board of directors may lose confidence in senior management, and the use of hostile consultants to find additional cost reductions to meet financial performance goals may be insisted on. None of these actions will reflect positively on senior management.

FAILURE TO MONITOR

Yet another pitfall hospitals experience is the failure to monitor the consistent efficiency of strategic cost reduction initiatives. Sometimes, even the best hospitals implement initiatives yet fail to monitor their permanency. It does a hospital no good, for example, to eliminate a staff position only to find out it increased the department's operating costs because more overtime is used several months after the staff elimination. Nevertheless, hospitals without sufficient monitoring systems to determine that strategic cost reductions actually take place and remain in place can sometimes face problems.

Retirement fiasco

A large urban hospital contemplated several strategies to reduce labor costs after determining that it was overstaffed by approximately 150

FTEs. Seeking to avoid a layoff, the CEO elected to offer an early retirement incentive program to encourage employees who were older than 55 years and had a tenure of 20 years or more to retire.

The good news was that approximately 140 employees elected to take early retirement, almost completely eliminating the overstaffing problem. The bad news was that no controls were placed on hiring new employees after the early retirement program was implemented. One year later, the hospital had hired nearly 200 new employees, creating an even worse overstaffing situation than before. It was subsequently discovered that the wisdom and experience of the retiring employees could not be easily replaced; in some cases, two new employees were hired to replace one experienced retiring employee.

Approximately $5 million of the hospital's funds were wasted on this retirement fiasco, and the CEO and vice president of human resources were terminated as a result of this failed strategy. It took nearly two more years for the hospital to reach its previous goal of reducing staff by 150 FTEs, and a layoff was ultimately necessary to reach this staffing goal.

For strategic cost reduction to be permanently effective, a hospital must avoid all of the aforementioned pitfalls. Failure to avoid them will ultimately mean failure to have strategic cost reduction fully impact the hospital. When that happens, the hospital's leadership must be held accountable. Alternatively, the hospital that identifies excellent strategic cost reduction initiatives will be in the best possible position to implement the initiatives and achieve the long-term value that strategic cost reduction can bring to the organization. The keys to delivering long-term success are careful monitoring, ensuring leadership accountability, and building a winning leadership culture, which are the subjects of the chapters in Part IV.

PART IV

Success Through Implementation

CHAPTERS 16 THROUGH 18 provide hospital CEOs and senior and middle management teams with the pathways to achieve and sustain strategic cost reduction success. Chapter 16 covers the important subject of monitoring strategic cost reduction initiatives after implementation. It highlights what to look for and potential problems that can occur. Chapter 17 discusses the topic of management accountability, clearly defining whose job it is to sustain strategic cost reduction success. Chapter 18 defines the elements of a winning corporate culture and the importance of culture on the future success of strategic cost reduction. The Afterword offers my insights on the inseparable connection among leadership, culture, and strategic cost reduction success.

Success with strategic cost reduction is not just a matter of effective implementation; the most detailed checklists and the most elaborate monitoring systems do not guarantee success. Success comes from the personal commitment of hospital leaders and their ongoing discipline to sustain green-dollar expense reductions without compromise. That level of leadership commitment and discipline can only come from personal accountability at all levels of the hospital, from employees to the CEO. When that level of accountability is present, the hospital has a winning leadership culture, which is the secret to success of strategic cost reduction.

CHAPT

Monitoring Strategic Cost Reductions

GENERATING A PLAN for green-dollar strategic cost reduction is the first, highly important step toward achieving success. Implementing the cost reduction ideas is the second critical step. Monitoring implementation for long-term success is the third and final step in the process. Many hospitals have succeeded with steps 1 and 2, only to fail with step 3. The positive effect of long-term strategic cost reductions cannot be sustained without accomplishing all three steps. To a hospital, the true value of implementing strategic cost reduction is reducing green-dollar costs on a permanent basis. This takes discipline and long-term commitment by the CEO and all members of management. Episodic cost reduction binges, regardless of how effective they may be in the short term, are not nearly as effective as long-term strategic cost reductions. Suggested approaches for monitoring strategic cost reduction initiatives follow.

MANAGEMENT COSTS

Because strategic cost reduction success begins with reducing management costs to the lowest possible level without compromising good leadership, these costs must receive the ongoing attention of the CEO and senior management team. When management layers have been reduced to a minimum and all ineffective members of management have been terminated, care must be taken to ensure that new management positions are not added unnecessarily. This takes vigilance and dedication, especially on the part of the CEO. The CEO must also be vigilant in ensuring that members of senior and middle management maintain high standards for the positive work ethic established during the strategic cost reduction process. Any slippage will immediately be observed by hospital physicians and employees and will send a mixed message about the commitment of senior management in maintaining strategic cost reduction success.

It is also recommended that an organizational chart be updated at least annually by the CEO and senior management. The typical hospital table of organization is presented on an 8½" × 11" standard sheet of paper, suitable for internal distribution. These little charts never reflect the true organizational structure and all of its management layers. The use of a boardroom table and a giant sheet of blank paper to draw the actual table of organization ensures that the CEO and senior management team continually update the layers of management between the CEO and front-line workers. By annually redrawing the true table of organization, hospital leadership can guard against adding layers (and management costs) after a strategic cost reduction initiative. In no case should the total number of management layers in a hospital ever be allowed to exceed six.

LABOR COSTS

Sustaining strategic reductions in the area of labor costs is perhaps the greatest of all management challenges. The pressures on hospitals to provide superb clinical quality and customer service are enormous. Management's natural tendency is to add rather than reduce staffing. Only the most disciplined hospital management teams can sustain reductions in labor costs while at the same time maintaining high quality care and excellent customer service for patients. Achieving strategic labor cost reductions begins with having a baseline labor-cost level for each department within the hospital. The use of a position control system is extraordinarily helpful in this area as the position control system specifies the number of authorized positions and the number of approved payroll hours of each position in a given cost center. In other words, the position control system provides a baseline on which future labor costs can be continuously measured.

Many complex and highly sophisticated systems are available for hospitals to measure labor costs and employee productivity. At its core, however, maintaining strategic cost reductions in the area of labor costs means maintaining a given level of paid hours per pay period on a continuous basis for each hospital department. If a department has an approved position control standard of 1,000 hours per pay period, any variance from that standard should be a cause for immediate follow-up. Hospitals that have the ability to monitor actual versus approved paid hours within 24 hours of the end of a payroll period have the best chance of succeeding with strategically managing labor costs in the long term. Within 24 hours of the end of a pay period, the director responsible for the department should receive a paid-hours report. The vice president responsible for that department should simultaneously receive the same payroll report for each department under her control. In this fashion, both directors and vice presidents can monitor the status of actual paid hours versus approved paid hours on a biweekly basis immediately after the end of a pay period. Again, any variances, regardless of the cause, should be immediately addressed.

Monitoring the use of outside agency staffing costs is also critical on a biweekly basis. During times of temporary or long-term staffing shortages, the use of agency labor has become commonplace in hospitals. Because agency staff typically costs two or three times as much as hospital employed staff, monitoring and managing agency labor expenses become a critical part of managing overall labor costs for hospitals.

Fringe-benefit costs need to be monitored with the same level of scrutiny as direct labor costs if strategic cost reductions for fringe-benefits are to be successfully lowered. Each hospital cost center should receive a report on actual fringe-benefit costs on a biweekly or at least monthly basis. Any variances from the approved baseline levels should be cause for immediate intervention. Controlling health insurance costs is especially important because this is typically the largest fringe-benefit cost hospitals experience.

SUPPLY COSTS

Each hospital department should establish an approved baseline level for supply costs after a strategic cost reduction initiative. Any expenditure of supply costs that is greater than the approved baseline level should be cause for immediate intervention, as with labor costs. Each strategic cost reduction initiative should be specifically identified and integrated into the baseline supply cost target. Written confirmation should be obtained from department directors that strategic supply cost reductions have been implemented in their respective areas of responsibility. After implementation, there should be a monthly review of any variances from the approved baseline supply-cost level. Variances in a department's volume should be taken into account when the baseline expenses are compared against the actual expenses. Increases in volume may be considered justification for proportional increases in supply costs. Conversely, decreases in volume should lead to a corresponding proportional reduction in supply costs.

SERVICE COSTS

As with supply costs, each hospital department should have a baseline level of approved green-dollar purchased service costs. These costs should be adjusted on the basis of the strategic cost reduction initiatives that are approved and implemented for each department. It is also appropriate for the responsible director or manager of each department to prepare written confirmation that strategic cost reduction initiatives have indeed been implemented.

Like labor and supply costs, variances in service costs from the approved levels should be cause for immediate intervention. The department's director is responsible for monitoring variances and responding appropriately. The hospital's vice presidents are responsible for ensuring that each director under their control also responds appropriately to variances in service costs from approved levels.

UTILIZATION COSTS

Clinical utilization costs must also be monitored extra carefully if the effect of strategic cost reduction initiatives are to be fully realized. Decreased costs because of decreased lengths of stay and lower clinical utilization rates should also be translated into approved baseline cost levels for each department within the hospital. Any variation in the approved utilization rates should, again, be cause for immediate intervention. Careful attention should be paid to changes related to volume and clinical utilization patterns attributable to increases or decreases in the case-mix index. Department directors and physician department chairs are responsible for monitoring clinical utilization costs and responding properly to variances from approved levels.

BUDGET PROCESS REFINEMENTS

Strategic cost reduction implementation should motivate hospitals to review their budgeting process and make refinements that integrate strategic cost reduction reasoning into the annual budgeting process. Strategic cost reduction success teaches the true meaning of green dollars to all levels within the hospital organizational structure. The concept of green dollars needs to be fully integrated into the budgeting process for directors and managers to be fully cognizant of all green dollars spent under their control. The budgeting process should provide historical information for directors and managers about labor, supply, service, and utilization costs. The baseline costs can then be analyzed carefully and integrated into future budget projections based on the green dollars that need to be spent to achieve the desired levels of quality and service.

In hospitals, it is rare for budgets to be used effectively at the department-director level for the purpose of monitoring actual-versus-budget performance on a month-to-month basis. Perhaps that is because, historically, directors and managers have not been as personally involved in building the budget as they could have been. With the strategic cost reduction process, department directors become intimately familiar with the green dollars expended for labor, supplies, services, and utilization in their areas of responsibility. This knowledge makes them far more able to contribute to building effective and accurate budgets that, in turn, can be used more effectively to monitor and control costs in the future.

CHECKLIST FOR MONITORING SUCCESS

Monitoring strategic cost reduction initiatives is absolutely essential to achieve long-term, positive results. The CEO, senior management, and middle management need to fully integrate the concept of green dollars into their day-to-day leadership activities. Following is a six-item checklist for monitoring long-term success:

1. *Implementation confirmation.* Each strategic cost reduction initiative should be implemented by the appropriate department director. This implementation should be confirmed, in writing, to the department director's supervisor in senior management.
2. *Baseline cost levels.* Each department should have a defined baseline level of labor costs, supply costs, service costs, and

clinical utilization costs after strategic cost reduction initiatives are implemented. These baseline levels should be incorporated into the hospital's operating budget.
3. *Labor cost monitoring.* Each pay period, actual labor costs should be immediately reported (within 24 hours) to each director. Any variances must also be immediately addressed to ensure absolute adherence to the baseline level.
4. *Other cost monitoring.* Each month, supply, service, and clinical utilization costs for each department should be promptly reported (within 5 days of month end) to every department within the hospital. Any variances must be immediately examined to ensure absolute adherence to the baseline level.
5. *Senior management.* Senior management must closely monitor all green-dollar costs for all departments and cost centers under their direct responsibility. Vice presidents should immediately address variances with their appropriate director. The CEO should monitor the hospital's overall green-dollar costs versus targeted levels and hold vice presidents accountable for achieving and sustaining the approved green-dollar cost reduction targets.
6. *Budgeting.* The concept of green dollars and the intimate understanding of labor, supply, service, and utilization costs should be used to improve the hospital's budgeting process. After a successful strategic cost reduction initiative, directors will be able to contribute to creating accurate cost budgets that are effectively used to monitor costs for the future.

These six checklist concepts take long-term commitment, discipline, and flexibility. Each concept needs to be taught and reinforced and, indeed, needs to be part of the hospital's leadership culture. Achieving strategic cost reduction takes an especially skilled management team to continuously monitor green-dollar costs. In other words, senior managers, middle managers, and department directors need to be held fully accountable for the green-dollar costs under their areas of responsibility. The importance of management accountability is the subject of Chapter 17.

CHAPTER 17

Accountability for Strategic Cost Reduction

STRATEGIC COST REDUCTION is the responsibility of everyone in the hospital. Strategic cost reduction can contribute significantly to achieving overall hospital financial performance goals. The CEO and senior management members are accountable for the hospital's financial performance. A positive leadership culture in the context of strategic cost reduction is one in which everyone, at all levels of the hospital, feels accountable for monitoring green-dollar costs. In this manner, every green-dollar expenditure is optimally efficient in achieving its intended result. As U.S. hospitals spend approximately $500 billion annually, accountability at all levels is one way to ensure that green dollars are spent wisely (Smith et al. 2006). A leadership culture that does not promote accountability at all levels cannot hope to achieve financial performance goals and success in strategic cost reduction. In those environments, no one is truly accountable for spending green dollars wisely.

SENIOR MANAGEMENT ACCOUNTABILITY

Accountability starts at the top. The CEO should personally set the example for a highly productive work ethic. A CEO's strong work ethic, combined with the expectation that the CEO's direct reports will also have a strong work ethic and high productivity, is the cornerstone of leadership accountability. Further, the CEO should strive to retain top performers as subordinates and should never accept the added expense of corporate dead weight for the reason that getting rid of ineffective executives is politically expedient or personally challenging. The CEO should also set a positive tone for the overall expenditure on supplies, services, and utilization. CEOs who are avidly interested in day-to-day hospital operations and can personally insist on the most effective and efficient expenditure of green dollars are likely to succeed with strategic cost reduction. CEOs who are distant from day-to-day operations, or who do not have the personal commitment to set high standards for productivity and other green-dollar costs, will never preside over a hospital that optimizes its financial performance.

As with the CEO, senior management also needs to set a positive example. Vice presidents should set a personal and positive example for all of their direct reports. They should work hard to set ideal work-ethic standards, and they should set a positive personal example for how supply and service funds are expended in their areas of responsibility. They should be knowledgeable enough about the operations in their departments to challenge directors and managers about green-dollar expenditures. They should be interested enough to know when to question labor, supply, service, and utilization costs. Further, vice presidents should never leave poorly performing managers or directors in place, regardless of how difficult it may be to find and train suitable replacements. By retaining ineffective performers, vice presidents set a poor leadership example and virtually guarantee that strategic cost reduction objectives will not be met in the departments led by these poorly performing directors.

MIDDLE MANAGEMENT ACCOUNTABILITY

As with the CEO and senior management, directors and managers are also accountable for strategic cost reduction success.

They must have a strong work ethic and individual productivity. Hard-working directors and managers set a positive example for employees. Directors and managers who lack personal productivity also set an example for employees: the wrong one. How can employees be inspired to work their hardest and produce the highest-quality work possible if their own supervisors are lacking in those areas?

> **Practical consolidation**
>
> A regional medical center engaged in strategic cost reduction carefully evaluated all middle management positions for redundancy. Part of this evaluation involved checking for overlapping responsibilities of department directors and identifying opportunities for consolidation. With input from vice presidents and department directors, it was determined that two director positions could be effectively combined into one. The director of pulmonary rehabilitation and director of cardiac rehabilitation positions were subsequently combined with no loss of leadership effectiveness in these clinical services. Because one director position and one secretary position were eliminated simultaneously, the green-dollar cost reduction exceeded $100,000.

Directors and managers must constantly focus on supply and service costs in their areas of responsibility. They are accountable for evaluating new technology, new protocols, and new procedures—all at the lowest possible costs. This, coupled with an emphasis on high employee productivity, guarantees that green dollars are being spent wisely. Although directors and managers are always looking for less expensive alternatives in the areas of labor costs, supplies, and services, they should be collaborating with physicians in the area of clinical utilization expenses. Directors and managers are much more closely associated with physicians and patients on a day-to-day basis than senior management is. Directors and managers who devote part of their energies to helping physicians be more efficient are promoting a culture in which strategic cost reduction can thrive.

PHYSICIAN ACCOUNTABILITY

Physicians, too, are accountable for supply and service expenditures, and they have a great deal of influence on labor costs. CEOs, senior managers, and middle managers all have an obligation to improve physicians' knowledge base concerning labor, supply, service, and clinical utilization costs. Physicians are rarely conscious of the green-dollar effect of their order-writing pen. How can they be expected to support a culture of strategic cost reduction when knowledge of these costs is limited? The answer does not lie in punitive programs or more intensive policing of physicians. The answer lies in increased knowledge.

Middle managers, especially, have the resources to familiarize physicians with the green-dollar costs of the supplies and services physicians order. Every director and manager should spend part of his time helping physicians understand the true green-dollar costs. If they do this personally, it is highly effective. Another highly effective strategy is for directors and managers to work intimately with physicians' support staff. Office managers and nurses associated with physicians' practices are usually very receptive to understanding the cost impact of the physicians' ordering patterns.

Appropriately armed with facts and figures, physicians are in a much better position to communicate effectively with their patients regarding hospital costs. Physicians should not be expected to make decisions wholly on the basis of costs. A physician who is in a position to explain all options to a patient and discuss the cost implications for those options is one who is enlisting the patient's preferences while helping to make sound clinical and economical decisions. This collaborative approach makes strategic cost reduction not only a hospital issue but also a patient customer issue.

EMPLOYEE ACCOUNTABILITY

Employees who feel personally accountable for green-dollar costs are powerful contributors to a hospital's achievement of strategic cost reduction goals. Employees represent those who provide nursing care, cook meals, take x-rays, clean floors, send bills, and perform a myriad of other tasks that in their entirety define hospital care. If at every point a green dollar is expended, an employee feels more personally accountable, it is a certainty that those dollars are spent more wisely. Where does this accountability begin?

It begins with employees who are qualified and motivated to provide quality care. Indeed, employees must have a good work ethic and a positive customer service attitude. Productive employees ensure that labor dollars are spent wisely. Unproductive, unqualified, or unenthusiastic employees just as certainly guarantee that labor dollars are spent unwisely. A constant flow of feedback to their direct supervisors on the efficacy of supply, service, and clinical utilization costs is part of the equation for achieving strategic cost reduction goals. A supply that is ineffective or does not achieve its intended use is a wasteful expenditure of green dollars. How else will supervisors gain insight into their departments unless employees communicate with them directly?

Employees can help physicians understand supply, service, and utilization costs. They are the physicians' representatives at the bedside. Employees who are part of a collaborative team provide feedback not only to their supervisors but also to physicians. Employee accountability can make or break a strategic cost reduction program.

PUTTING IT ALL TOGETHER

Successful strategic cost reduction requires a leadership culture of accountability. An excellent leadership culture will always have the following key characteristics:

- Positive leadership example
- Intense focus on the basics
- Excellent communication
- Constant adjustments

These are the same basic principles of clinical and customer service excellence. Strategic cost reduction is a cornerstone to an excellently performing hospital, as are clinical excellence and service excellence. All three must be present for a hospital to truly achieve optimal performance. The absence of any one of these three will make greatness impossible to achieve. On the other hand, the presence of all three elements does not guarantee success. Only excellent leadership can guarantee sustained success for a hospital. Leadership excellence takes a winning corporate culture to thrive, and this is the subject of Chapter 18.

Reference

Smith, C., C. Cowan, S. Heffler, A. Caitlin. 2006. "National Health Spending in 2004: Recent Slowdown Led by Prescription Drug Spending." *Health Affairs* (Jan/Feb): 186–96.

CHAPTER 18

Building a Winning Strategic Cost Reduction Culture

SUCCESSFUL STRATEGIC COST reduction is a leadership challenge, much more so than a financial challenge. One of the key ingredients for sustained success is a winning leadership culture that supports and enhances strategic cost reduction actions. Some might argue that the presence of a positive leadership culture is a necessary ingredient for attaining strategic cost reduction success. In my experience, however, building a winning leadership culture is part of the process of achieving strategic cost reduction success. One does not precede the other; they are accomplished simultaneously. For some hospitals, creating this kind of leadership culture is the ultimate benefit of strategic cost reduction.

Plenty of examples show low cost and high quality co-existing successfully. In the airline industry, Southwest Airlines and Jet Blue Airways have a record of low-cost structures combined with superb quality service. In an industry where red ink gushes, these two airlines have a solid history of profitability. Both have strong corporate cultures that emphasize spending green dollars wisely, maintaining extraordinarily high employee productivity, and providing exceptional customer service. The hospital industry would do well to emulate their success. Strategic cost reduction is one important methodology to accomplish this.

Leadership culture permeates the hospital from top to bottom. It guides decision making at all levels and defines the behavior within a hospital. A winning leadership culture in the context of strategic cost reduction begins with trust and accountability.

TRUST AND ACCOUNTABILITY

One of the basic principles of strategic cost reduction is that middle managers, physicians, and employees know best where green-dollar costs can be permanently reduced without harming quality of care or quality of customer service. This basic premise underscores the trust that senior management must place in middle management to make decisions to reduce costs without doing harm. Trust must emanate from the CEO and senior management. If trust is present, and if members of middle management are confident in their personal leadership capabilities, strategic cost reduction can succeed.

This level of trust also emanates from middle managers to the physicians and employees with whom they interact on a day-to-day basis. If middle managers trust that their employees will provide good insights and suggestions regarding the permanent reduction of costs, it shows that they trust their own employees as much as senior management trusts them. The same can be said for physicians. If middle managers seek and listen to suggestions from physicians they interact with and act on those suggestions, this too demonstrates a level of trust that becomes part of the hospital's culture.

With trust comes accountability. When middle managers are held accountable for the quality of care for services they provide, they, in turn, will also hold the physicians and employees accountable. In this manner, trust and accountability go hand in hand to create a cultural environment that enables strategic cost reduction ideas to flourish, be implemented, and be sustained on a long-term basis.

SUSTAINING STRATEGIC COST REDUCTIONS

Most hospitals can develop ideas for reducing costs, either on their own or by using consultants, but it takes an extraordinary hospital leadership and a winning leadership culture to implement and sustain those strategic cost reduction initiatives. Implementing and sustaining strategic cost reductions is much more difficult than developing the initial ideas. Only a winning leadership culture that promotes trust and accountability has the best chance of succeeding.

Sustaining labor-cost reductions requires especially good leadership among middle management ranks. It requires hiring the best and most competent employees, setting high expectations for quality and customer service, and maintaining those standards over time. It also requires leaders to excel in times when employee performance falls short. In those instances, middle management leaders must either address performance problems successfully or have the discipline and courage to dismiss the weakest employees and find those who are more capable. A leadership culture that allows, and even demands, middle managers to function under these basic leadership guidelines is an absolute necessity for strategic cost reduction success.

The same kind of discipline required for middle managers is also required for sustaining cost reductions in supplies and services on an ongoing, permanent basis. Leadership culture must instill in middle managers the discipline to constantly evaluate the efficacy of their supply and service decisions according to strategic cost reduction principles. The discipline and insight needed to constantly reevaluate decisions are never easy; constant evaluation requires self-awareness and self-critiquing skills that not everyone possesses. Assuredly, the leadership culture that cherishes those skills has the best chance of succeeding with strategic cost reduction initiatives on an ongoing basis.

SUPERB COMMUNICATION

Another aspect of a winning leadership culture is superb and frequent communication. Senior management, beginning with the CEO, must set this positive example. The CEO must promote the hospital's vision, the hospital's strategic objectives, and the case for strategic cost reduction. He must introduce his personal leadership expectations for senior and middle management as well as for physicians and employees. Next, the CEO should ensure that frequent status reports are part of the hospital's communication culture. Status reports must be complete, honest, and encouraging.

Middle managers must also be expert communicators. It is their job to interpret the hospital's vision and expectations of employees that will help the hospital achieve its strategic goals. It is also their job to continuously solicit and listen to feedback from employees on how the strategic initiatives and the hospital's vision can be enacted.

Finally, employees must practice vigorous communication skills in a winning leadership culture that achieves strategic cost reduction goals. They must communicate with physicians, with their fellow employees, and with customers. They must be good listeners and providers of feedback. They must be respectful toward hospital leaders and patients. Almost certainly, the absence of good communication at the employee level will inhibit even the most positive leadership culture from thriving and succeeding in strategic cost reduction.

LEADERSHIP TRAINING

A winning leadership culture cherishes teaching and training. The CEO should be thought of as the principal leadership teacher for the hospital, the one who sets the overall leadership tone. But, the CEO cannot be the sole teacher; a winning leadership culture is one that instills the teaching obligation at all levels of management, even among physicians. It is incumbent on each member of management to teach and inspire those who report to them. It is physicians' obligation to teach and inspire patients so that compliance with their clinical and lifestyle recommendations is met. Winning leadership cultures are always looking for opportunities to teach and perfect, elusive though they may be.

POSITIVE REINFORCEMENT

A winning leadership culture is one that values positive reinforcement much more than negative reinforcement. All levels of management, beginning with the CEO, should seek opportunities to positively reinforce any individual who reports to them. In this manner, employees will experience the benefits of a positive culture, rather than experience constant critiques that can be commonplace in negatively oriented cultures.

Many opinions have been written about corporate cultures. What is past dispute, however, is that a winning hospital leadership culture has four "bests": the best leaders, the best communication, the best quality of care, and the best customer service. Achieving strategic cost reduction success is ultimately what enables hospitals to sustain the four bests.

Afterword

I HAVE HAD the opportunity to work with some of the most successful hospital leadership teams and some of the most unsuccessful hospital leadership teams. The best leadership teams were those that retained my company to assist them with strategic cost reduction. The worst leadership teams were those that hired my company to lead financial turnarounds for their hospitals after experiencing years of poor financial performance.

Surely, lessons are to be learned from the best and worst leadership teams. The best CEOs always seem to be personally involved with both strategy and hospital operations. They participate in strategy and vision formation while remaining intimately interested in the hospital's day-to-day operations. The best CEOs never view themselves as outside leaders. They embrace a leadership culture that reflects a deep appreciation for quality of care and customer service; they are truly engaged leaders.

The worst CEOs always seem to display the opposite traits. They are uninvolved. They appear to be more interested in outside activities than in the day-to-day operations of the hospital. The worst leaders have the habit of rationalizing poor financial and clinical performance. In other words, it is never their fault. The blame for failure always seems to be an external dark force, such as the government, managed care, or an aggressive competitor. They never seem to look into the mirror when seeking to determine who is at fault when the hospital fails to thrive.

Strategic cost reduction is one of the answers to this nation's healthcare crises. Hospitals could reduce their total costs by 5 percent or more and do no harm. If this occurred, the savings could begin to solve the problem of our country's uninsured. The best CEOs and their senior management teams embrace strategic cost reduction; they succeed in lowering green-dollar costs within their hospitals without diminishing clinical quality and customer service. The worst leadership teams continue placing the blame for their hospital's failure everywhere but where it really belongs—themselves. I look forward to meeting more of the best leadership teams in the exciting years that lie ahead.

APPENDIX A

Background Information Trend Guide

Anytown Memorial Hospital

Financial Trends: Fiscal Years 2003–2007

- Total patient services revenue
- Total deductions from revenue
- Net patient services revenue
- Total expenses
- Income (loss) from operations
- Budgeted income (loss) from operations
- Accounts receivable days
- Accounts payable days
- Days cash on hand
- Average age of plant and equipment

Activity Trends: Fiscal Years 2003–2007

- Inpatient admissions
- Average daily census
- Obstetric deliveries

- Emergency department visits
- Outpatient visits
- Inpatient surgery cases
- Outpatient surgery cases
- Overall length of stay
- Medicare length of stay
- Total FTEs
- FTEs per adjusted occupied bed

Internal and External Reports

- Most recent audited financial statement and management letter
- Fiscal year 2008 operating and capital budgets
- Pertinent, internally developed or consultant-assisted financial strategic cost reduction reports (for the past 36 months)
- Most recent market research, patient satisfaction, physician satisfaction, and employee satisfaction survey results

APPENDIX B

Sample Calendar for Strategic Cost Reduction

Anytown Memorial Hospital April through August 200X

Wednesday, April 20, 200X

9:00 a.m.	Director/manager meeting Introduction: Strategic adviser
11:00 a.m.	Administrative staff meeting Introduction: Strategic adviser
1:00 p.m.	Director/manager focus group*
3:00 p.m.	Director/manager focus group

Thursday, April 21, 200X

9:00 a.m.	Director/manager focus group
11:00 a.m.	Director/manager focus group
2:00 p.m.	Director/manager focus group

Tuesday, April 26, 200X

11:00 a.m.	Director/manager focus group

*Focus groups conducted by the strategic adviser to introduce strategic cost reduction concepts and create an opportunity for the strategic adviser to learn about Anytown Memorial Hospital's corporate culture.

1:00 p.m.   Director/manager focus group
2:30 p.m.   Director/manager focus group

Monday, May 2, 200X

2:00 p.m.   Administrative staff: Strategic cost reduction planning

Tuesday, May 3, 200X

9:00 a.m.   Kickoff retreat: Strategic cost reduction

Monday, May 9, 200X

11:00 a.m.   Task force co-chairs' meeting with strategic adviser

Monday, May 16, 200X (Meeting with strategic adviser)

12:30 p.m.   Task force 1
2:00 p.m.   Task force 2
3:30 p.m.   Task force 3

Tuesday, May 17, 200X (Meeting with strategic adviser)

8:30 a.m.   Task force 4
10:00 a.m.   Task force 5
11:30 a.m.   Task force 6
1:00 p.m.   Task force 7
2:30 p.m.   Task force 8
4:00 p.m.   Task force 9

Monday, May 23, 200X

11:00 a.m.   Task force co-chairs' meeting with strategic adviser

Monday, June 6, 200X (Meeting with strategic adviser)

12:30 p.m.   Task force 1
2:00 p.m.   Task force 2
3:30 p.m.   Task force 3

Tuesday, June 7, 200X (Meeting with strategic adviser)

8:30 a.m.	Task force 4
10:00 a.m.	Task force 5
11:30 a.m.	Task force 6
1:00 p.m.	Task force 7
2:30 p.m.	Task force 8
4:00 p.m.	Task force 9

Monday, June 20, 200X (Meeting with strategic adviser)

11:00 a.m.	Task force co-chairs
12:30 p.m.	Task force 1
2:00 p.m.	Task force 2
3:30 p.m.	Task force 3

Tuesday, June 21, 200X (Meeting with strategic adviser)

8:30 a.m.	Task force 4
10:00 a.m.	Task force 5
11:30 a.m.	Task force 6
1:00 p.m.	Task force 7
2:30 p.m.	Task force 8
4:00 p.m.	Task force 9

Wednesday, June 22, 200X

8:30 a.m.	Director/manager meeting Strategic cost reduction update

Thursday, June 30, 200X

11:00 a.m.	Task force co-chairs' meeting with strategic adviser

Monday, July 11, 200X (Meeting with strategic adviser)

12:30 p.m.	Task force 1

2:00 p.m.   Task force 2
3:30 p.m.   Task force 3

Tuesday, July 12, 200X (Meeting with strategic adviser)

8:30 a.m.   Task force 4
10:00 a.m.   Task force 5
11:30 a.m.   Task force 6
1:00 p.m.   Task force 7
2:30 p.m.   Task force 8
4:00 p.m.   Task force 9

Wednesday, July 20, 200X

11:00 a.m.   Task force co-chairs' meeting with strategic adviser

Tuesday, July 26, 200X (Meeting with strategic adviser)

12:30 p.m.   Task force 1
2:00 p.m.   Task force 2
3:30 p.m.   Task force 3

Wednesday, July 27, 200X (Meeting with strategic adviser)

8:30 a.m.   Task force 4
10:00 a.m.   Task force 5
11:30 a.m.   Task force 6
1:00 p.m.   Task force 7
2:30 p.m.   Task force 8
4:00 p.m.   Task force 9

Thursday, July 28, 200X

9:00 a.m. – 12:00 p.m.   Administrative staff: Review task force recommendations

Thursday, August 11, 200X

8:00 a.m. – 11:00 a.m.	Administrative staff: Review task force recommendations
11:00 a.m.	Task force co-chairs' meeting with strategic adviser

Saturday, August 13, 200X

9:00 a.m. – 2:00 p.m.	Administrative staff: Final review of strategic cost reduction plan

Wednesday, August 17, 200X

9:00 a.m. – 12:00 p.m.	Administrative staff: Implementation planning for strategic cost reduction initiatives

Monday, August 22, 200X

4:00 p.m.	Board finance committee: Review implementation plans

Tuesday, August 23, 200X

5:00 p.m.	Board of directors' approval of implementation plans

Thursday, August 25, 200X

3:00 p.m.	Leadership celebration retreat for board of directors, administration, and all task force members

Monday, August 29, 200X

Strategic cost reduction implementation

APPENDIX C

Sample Strategic Cost Reduction Scenario

Anytown Memorial Hospital

ANYTOWN MEMORIAL HOSPITAL has served our community with great success for more than 75 years. Despite our history, however, our hospital faces many challenges in the future. Among our most significant challenges are replacing our aging 1970 inpatient tower, updating critical care and operating room facilities, and expanding the emergency department to accommodate our growing patient population.

We must meet these challenges at a time when the state government is reducing Medicaid payments next year by 5 percent and the federal government is reducing Medicare payments by 2 percent. How can we simultaneously meet local healthcare challenges while adapting to reimbursement reductions at state and federal levels? We will show extraordinary leadership by strategically reducing costs we directly control. We will evaluate every dollar we spend on staffing, supplies, and services and make reductions wherever possible based on the collective wisdom of our management team, physicians, and employees.

Anytown Memorial Hospital has set a goal of reducing operating costs by $20 million per year. We will reduce these costs while maintaining and improving customer service and patient care quality. Our success will ensure that Anytown Memorial Hospital has the necessary resources to succeed in the future as we have in the past.

APPENDIX D

Sample Town Meeting Questions and Answers

May 200X

FOLLOWING ARE QUESTIONS commonly asked during strategic cost reduction initiatives. Suggested answers are given to help clarify the intent of this project.

Q. How was the overall goal of $20 million set, and how are the goals for the individual task forces established?

A. The overall goal was the minimum savings necessary for Anytown Memorial Hospital (AMH) to achieve a level of financial performance in fiscal year 2008 that is consistent with funding AMH strategic goals for the future. Individual task force goals were determined by apportioning the overall goal fairly to task forces according to the operating expenses that are directly controlled by task force members.

Q. Why is the development of new facilities required at this time?

A. AMH is operating at full capacity in many departments. With our growing market and ever-increasing demand for our services, we risk losing current and future business if we do not enhance and expand our facilities.

Q. How were task force members selected?

A. Nine task forces were created, and they are composed of 100 managers, directors, and vice presidents. All members of the AMH management team who have direct day-to-day control of budgets will participate in task forces. Task force members are expected to gain input from all employees, physicians, and managers who are not serving on task forces in their areas of responsibility.

Q. Who leads the task forces?

A. Co-chairs will be selected by members of each task force after the kickoff retreat on May 3, 200X.

Q. Will task forces receive additional support to enhance their chance of success?

A. Yes. Each task force will have a support staff person from finance and human resources to help task force members research ideas to reduce operating expenses.

Q. What can we tell employees about strategic cost reduction?

A. Task forces are gathering and brainstorming potential cost reduction ideas. Explain that this is work in progress and not a finished product at this time. All employees should be encouraged to share their ideas for cost reduction. Explain that cost reduction changes can be achieved without compromising patient care quality.

Q. What can we tell physicians about strategic cost reduction?

A. Task forces are gathering and brainstorming potential cost reduction ideas. Involve physicians actively in deliberations and invite their participation. Explain that this is a work in progress and that cost reduction changes can be achieved without compromising patient care quality. Also, be assured that members of administration will be having dialogues with physicians to enlist their support and solicit ideas for reducing costs while preserving high quality and service to patients.

Q. Can we make our best assumptions as we redesign and reorganize our departments for the future?

A. Yes. You may establish more than one scenario for a particular recommendation. Be sure to include documentation of the assumptions you have followed for each idea.

Q. Is there a cutoff date to include past staff savings toward the goal?

A. Consider March 1, 200X, as the cutoff date for past staff changes. For example, if a position was vacated on or after March 1, 200X, and you are not going to fill that position, you can include the dollars associated with that position toward your task force's goal. If the position was vacated before that date, do not include the savings.

Q. Can we include fringe-benefits savings as part of our goal?

A. FICA savings (7.65 percent) can be included as green-dollar savings. These savings should be identified and totaled separately from wages/salaries. No other fringe benefits can be included as part of the group goal as the administrative staff will consider them separately.

Q. What time period should the savings cover?

A. We are looking for permanent savings that will occur beginning October 1, 200X, or before. If you can identify savings that will occur in later years, list these amounts but do not include them as part of your goal.

Q. What happens if a cost savings initiative is developed that requires buy in from members of two different task forces to implement?

A. The savings will be divided equally between both task forces.

Q. Should task forces consider cost savings opportunities that could require capital investments?

A. Yes. All cost savings opportunities should be fully explored. Estimates of capital costs associated with a cost reduction strategy should be provided, together with a detailed explanation of the operations savings that could be achieved.

APPENDIX E

Sample Task Force Guidelines

Anytown Memorial Hospital

THE FOLLOWING GUIDELINES and suggestions are intended to help Anytown Memorial Hospital's strategic cost reduction task forces to get started.

1. *Goal and baseline.* The overall cost reduction goal is $20 million from the baseline of current expenses. This goal has been divided among the nine task forces according to the operating expenses directly controlled by members of each task force. Each task force will receive confirmation of its specific cost reduction goal on May 9, 200X at the first meeting of the task force co-chairs.
2. *Task force leadership.* Task force team members will be announced at the kickoff retreat on May 3, 200X. Each task force will select two co-chairs and provide their names to the president no later than noon on Friday, May 6, 200X. Co-chairs will share the leadership duties throughout the strategic performance improvement project. An in-service training session on meeting facilitation will be provided at the first meeting of co-chairs, on May 9, 200X, at 11:00 a.m.

3. *Cost savings opportunities.* Task force members should explore all areas of costs. Staffing, supplies, purchased services, and programs should be reviewed for potential savings opportunities. There are no sacred cows. All potential cost savings will be given serious consideration, even those that may have been turned down in past years. To be counted toward a task force's goal, cost savings must be actual green-dollar savings as defined below.
4. *Green dollars.* To qualify as cost reductions, all savings must be in green dollars. Green-dollar savings are defined as actual reductions in current expenses. In other words, savings must result in fewer dollars paid out in the future for supplies, services, or staff than are paid in the present. Future checks audits will be performed to verify that the savings task force members propose have actually occurred.
5. *Task force support.* Each task force will have a representative from finance and from human resources to help task force members research cost reduction initiatives.
6. *Staffing expenses.* To achieve the overall savings goal of $20 million, some staffing reductions will likely be necessary. If a task force identifies a cost savings opportunity through staff reductions, it will be asked how quality will be affected by the recommendation. Permanent decisions not to fill positions vacated through turnover after March 1, 200X will be counted as savings toward the task force's goal.
7. *Participation.* Each member of the task force must identify savings she is personally willing to implement and for which she has support to implement within her area of responsibility. There will be no across-the-board reductions. The level of cost savings in each department will be agreed on by the entire task force. It is up to the co-chairs and all members of the task force to ensure that cost reductions are applied fairly.
8. *Revenue.* Although revenue is not the initial focus of the strategic performance improvement project, task forces are encouraged to identify revenue opportunities for fiscal year 200X and beyond.

However, these dollars may not be used to offset expense reductions for purposes of achieving the task force's goal.
9. *Task force meetings.* Each task force is expected to meet at least two to three times per week through July 29, 200X. In addition, progress meetings will also be held with the strategic adviser approximately every two weeks.
10. *Communication.* Open and honest communication and a positive attitude about the strategic performance improvement project are expected. Input from employees and physicians is necessary and will be essential to achieve the overall goal. It is expected that no one will disclose information that task forces consider to be confidential.
11. *Reporting cost reduction ideas.* A sample report format will be provided for all task forces to identify cost reduction strategies consistently.
12. *Weekly progress reports.* Each task force is expected to prepare a weekly progress report for the president. Progress reports should identify the number of meetings held, best ideas of the week, problem areas encountered, and the total dollar value of cost savings identified to date. Weekly progress reports are due in the president's office no later than noon each Friday, beginning Friday, May 20, 200X, and continuing throughout the strategic cost reduction process.
13. *Final report.* Each task force will provide its final report of cost reduction recommendations to the president's office on or before noon on Friday, July 29, 200X. Each task force is expected to identify cost savings that are equal to or greater than their assigned goal.
14. *Questions and problems.* Task forces may use their co-chairs as the first resource for questions and problems. The strategic adviser, members of the administrative staff, and the finance and human resources support staff assigned to each task force are also available to help.
15. *Customer service and employee morale.* Throughout the strategic cost reduction process, task forces will be challenged to

identify initiatives to improve customer service and employee and physician morale. The creative and leadership challenge for task forces will be to identify service and morale improvement opportunities that can be implemented simultaneously with cost reduction initiatives.

16. *Demonstrated leadership.* Each task force member must consistently demonstrate positive commitment to the strategic cost reduction process. Task force members will be highly visible, and physicians and employees will be highly observant of their attitudes and actions throughout this process. It is very important for task force members to reinforce for all internal and external constituencies that they are personally making the recommendations necessary to support the revitalization and future growth needs of Anytown Memorial Hospital.

APPENDIX F

Sample Task Force Meeting Guidelines

Anytown Memorial Hospital

1. Develop meeting calendar for entire project.
2. Start meetings on time, and use short agendas.
3. Add items to agenda at beginning of meetings.
4. Keep meetings flowing.
5. Keep speakers in order.
6. Invite guests as appropriate.
7. Avoid "sidebar" conversations.
8. Solicit agenda items for the next meeting at the end of the meeting.
9. Maintain confidentiality of all task force discussions.
10. Ask for help from administration or the strategic adviser as necessary.

APPENDIX G

Sample Weekly Task Force Report

Anytown Memorial Hospital

To: John Q. Smith, President
Anytown Memorial Hospital
From: Mary Jones, co-chair of Task Force 6
Date: June 11, 200X

Task force 6 met three times this week, on June 3rd, 6th, and 10th. All task force members attended, and meetings lasted approximately 90 minutes. Following are the key topics of discussion this week:

Two supervisory positions could potentially be combined in the radiology department because of the upcoming retirement of the supervisor of computed tomography. We are exploring the possibility of combining the computed tomography supervisor and magnetic resonance imaging supervisor positions, which would create green-dollar savings of $72,000 annually.

The operating room director is engaged in a competitive bidding update for suture supplies with the help of the chief of the surgery department. Three vendors are under consideration, and potential annual savings exceed $100,000. Support from administration for a change in vendors will be essential.

The nursing director of medical surgical 3 west is evaluating a staffing-mix change on the 11:00 p.m. to 7:00 a.m. shift. This evaluation will be completed by next week's report.

We are concerned that potential morale problems are being created in the laboratory department because of rumors of a pending layoff. Would it be possible for a member of administration to meet with the laboratory staff next week?

Task force 6 has identified $450,500 in green-dollar savings initiatives to-date. This is 48 percent of our assigned goal of $925,000. We are confident that our task force will meet its goal by the end of the strategic cost reduction project. Please call or e-mail if you have any questions.

APPENDIX H

Sample Board of Directors Report

Anytown Memorial Hospital

Strategic Cost Reduction Timeline

May 200X	Strategic cost reduction kickoff meeting
June/July 200X	Task force recommendations
August 200X	Administrative staff review/approval
	Approval of reductions by the board of directors
September 200X	Implementation strategic cost reductions

Strategic Cost Reduction Goal

Overall goal	$20 million
Actual result	$22+ million

Cost Reduction Synopsis

Labor savings	54%
Supply reductions	22%
Purchased service savings	26%
Total	100%

Effect on Staffing

	FTE Reductions
Management staff	28
Agency/contract staff	20
Attrition and retirement	61
Overtime elimination	23
Reduction in hours	29
Positions eliminated	47
Other	23
Total staffing reductions	231

Management Accountability Improvements

- New performance review process in place
- New pay-increase standards
- Retention versus separation decisions based on performance
- Emphasis on future leadership development

Strategic Cost Reduction Synopsis

- Bottom-up process
- Management, physician, and employee participation
- Goal of $20 million exceeded
- Organizational leadership much stronger
- Administration ready for implementation

Board Motion to Approve Strategic Cost Reduction Plan

APPENDIX I

Sample Detailed Cost Reduction

Anytown Memorial Hospital
Strategic Performance Improvement

AT THE CONCLUSION of a strategic cost reduction initiative, a detailed final report should be prepared to identify each cost reduction initiative. This report should specify the department and include a brief description of the cost reduction initiative, the current cost, and the intended cost reduction impact of the initiative. It should also specify the FTE effect of the initiative, the intended implementation date, and the person responsible for implementation. Detailed cost reduction reports can later be used for follow-up audits to ensure that implementation has been successfully accomplished and that the intended financial impact has been realized.

Dept. Name	Description	Current Cost	Annual Savings	FTE Effect	Start Date	Responsibility	Comments/Effects
Supply, processing, and distribution	Renegotiate contract for sequential stockings from $32 per unit to $18 per unit	$121,201	$53,000	0.00	10/1/0X	N. Brown	Better pricing achieved through negotiation
Respiratory therapy	Renegotiate medical director's fees	$429,000	$42,000	0.00	9/1/0X	S. Jones	Contract renegotiated in the Sleep Center for medical director fees on 7/14/0X
Laboratory	Conversion from purchased analyzer to lease arrangement will eliminate service contracts on the South Campus	$26,000	$26,000	N/A	9/1/0X	T. Johnson	Savings resulted from the new contractual arrangement with Dade for general chemistry; with that agreement, two chemistry analyzers on the main campus will no longer require service contracts
Public relations	Communicare, a health education publication produced by the public relations staff, will be published biannually rather than quarterly; also, use of a new printing vendor will enable us to reduce print costs	$80,866	$50,000	N/A	10/1/0X	C. White	Health education information will be published instead in the bimonthly *Living* magazine; expect to increase frequency of release of information to the community through use this publication
Electrocardiogram	Eliminate coordinator position	$48,442	$48,442	1.00	10/1/0X	T. Washington	Reduction of 1 FTE; coordinator position is not justified with new department organization
Employee health	Discontinue providing annual blood work to employees	$28,000	$28,000	0.00	10/1/0X	B. Doe	Employees can get laboratory work done at their private physician's office under their copayment plan; in some cases, employees are receiving duplicate laboratory work from Anytown Memorial Hospital and the physicians' offices
Provider relations	Eliminate quality improvement specialist FTE	$58,400	$58,400	1.00	Immediate	W. Thompson	Eliminate position; redistribute duties and responsibilities of this position among four staff members

About the Author

Michael E. Rindler is an accomplished adviser, CEO, and author with three decades of experience assisting hospitals and healthcare systems throughout the United States. After 15 years of leading community hospitals and a regional medical center, he founded The Rindler Group in 1989, followed by Integrity Hospital Company in 2003.

Currently, Mr. Rindler serves as president and CEO of Integrity Hospital Company, a healthcare consulting and management company. Mr. Rindler advises hospitals and healthcare systems on strategic cost reduction, governance restructuring, leadership performance improvement, and strategic planning. As an interim CEO, he has also personally led ten successful hospital and health system turnarounds since 1989.

Mr. Rindler is an important contributor to the healthcare industry. He has previously authored four books: *The Essential Guide to Managing Consultants: Strategies for Healthcare Leaders; The Challenge of Hospital Governance: How to Become an Exemplary Board; Managing a Hospital Turnaround: From Crisis to Profitability in Three Challenging Years;* and *Putting Patients and Profits into Perspective.* These books provide valuable insights on healthcare consulting, hospital governance, and executive leadership.

Mr. Rindler holds a master of management degree from Northwestern University's Kellogg School of management and a bachelor of science degree from the University of Illinois. He enjoys flying vintage airplanes and collecting vintage cars. Mr. Rindler's marine and automobile photographs are owned by private collectors throughout the United States. He resides in Southwest Harbor, Maine, and Charleston, South Carolina, with his family.